CW01197724

MANIFESTING WITH VIBRATIONS

Learn the secrets with the 30-day plan to increase your vibrations, success in your life, freedom and lasting happiness, awareness, abundance, and attunement to the Universe

Samantha Goleman

TABLE OF CONTENTS

Introduction 4
What are Vibrations? 10
Vibrations of the Universe 23
    Distinguishing Learning from Inspiration 23
    Feel the Universe in Meditation 24
Fourth Dimension 29
Vibration Profile 32
Determining Your Vibration Profile 34
Attuning to the Vibration of the Universe 36
Mindfulness 41
Strength from Within 50
Manifesting 57
    Rewards vs. Achievement 59
Where Manifestation Comes From 61
How to Manifest 66
    Clearing 66
    Feel Your Intention 71
    Take Inspired Action 73
How to Raise Your Vibration 74
    Rituals 75
    Scripting 76
Creative Thinking to Improve Your Success 78
Living an Abundant Life 81
    Shift Your Focus Regarding Bills and Debts 82

Realize How Much Abundance There Is Already in Your Life 83

Love Money 84

15 Best Secrets of Manifestation 86

True Freedom and Lasting Happiness 97

30-Day Plan to Raise Your Vibration 106

Conclusion 115

Text Copyright © [SAMANTHA GOLEMAN]

All rights reserved. No part of this guide may be reproduced in any form without permission in writing from the publisher except in the case of brief quotations embodied in critical articles or reviews.

Legal & Disclaimer

The information contained in this book and its contents is not designed to replace or take the place of any form of medical or professional advice; and is not meant to replace the need for independent medical, financial, legal or other professional advice or services, as may be required. The content and information in this book has been provided for educational and entertainment purposes only.

The content and information contained in this book has been compiled from sources deemed reliable, and it is accurate to the best of the Author's knowledge, information and belief. However, the Author cannot guarantee its accuracy and validity and cannot be held liable for any errors and/or omissions. Further, changes are periodically made to this book as and when needed. Where appropriate and/or necessary, you must consult a professional (including but not limited to your doctor, attorney, financial advisor or such other professional advisor) before using any of the suggested remedies, techniques, or information in this book.

Upon using the contents and information contained in this book, you agree to hold harmless the Author from and against any damages, costs, and expenses, including any legal fees potentially resulting from the application of any of the information provided by this book. This disclaimer applies to any loss, damages or injury caused by the use and application, whether directly or indirectly, of any advice or information presented, whether for breach of contract, tort, negligence, personal injury, criminal intent, or under any other cause of action.

You agree to accept all risks of using the information presented inside this book.

You agree that by continuing to read this book, where appropriate and/or necessary, you shall consult a professional (including but not limited to your doctor, attorney, or financial advisor or such other advisor as needed) before using any of the suggested remedies, techniques, or information in this book

.

Introduction

The Universe is a part of all of us. Some say that the energy that is part of everything is God and God is everything. If so, we are all a part of God and we have the power to make things happen more than we thought possible. The power to manifest – to make tangible inspirations that are intangible is certainly divine in nature. Remove your blinders and your limiting beliefs and you will come to understand the true nature of yourself, this world, and the Universe in and around you.

At its most fundamental level, this Universe is made up of energy. From this energy, comes matter. From matter, come the various particles, and from particles we get the elements. If you take hydrogen, you find that it is made up of one electron and one proton – two different particles. If you added another particle to that existing hydrogen atom, you get helium. And so you get one element after another. But the starting point is the same, energy.

Einstein figured this out when he derived the *e=mc2* equation. You've seen it everywhere; you see it all the time. You probably even realize that it means that mass and energy are interchangeable. But it is also more than that. Matter and energy are the same things in different states. It is kind of like water steam and ice are the same thing, only in different states.

From a superficial perspective, ice is tangible; you can hold it, observe it with your senses and even count it.

On the other hand steam (in the absence of equipment) cannot be seen, touched, or measured with your bare senses. So for this discussion let's just take that this water steam is intangible.

Energy and matter have the same parallel. Matter is tangible and energy is not (again, in the absence of special equipment). Your body is the same. It is composed of tissue and soul. Tissue is tangible and the soul is not.

This world is built on tangible and intangible. The tangible are the parts you can observe while the intangible are the parts you cannot observe directly but you know they are here because of the way other elements or matter behave around them.

What does all this got have to do with meditation?

Well just as sound is tangible, silence is intangible. So if we wanted a nice little package, here is what we have – matter, water, and the tissue of the body can be placed under the tangible column; while energy, steam and soul are placed in the intangible column.

Sounds are tangible because you can use one of your senses to detect them. Feelings are tangible too because feelings are merely chemical reactions in your body that result in a certain level of detection. But silence is undetectable by any sense in your body and only discernible by exclusion. For this reason, silence is considered intangible.

Energy, the soul, and silence are in the same column since they are intangible. We consider the soul to be part of the universal energy that is within us. So when you want to understand the soul and the Universe by extension, you have to use silence. But that is a lot easier than it sounds. It's not sufficient to stop talking or stop listening. Invoking this type of silence requires a different method, and that is the method of meditation.

The distractions of the mind that are constantly battling with the body are distractions that overshadow the silence of the soul. To reach that silence, you have to silence the chaos of the mind. This is the reason most people misunderstand meditation to mean the quieting of the mind. You have to still the mind before being able to appreciate meditation.

This is the key to manifestation. You need a clear mind. Once you do that, your ability to vibrate your desires improves exponentially. Your ability to attract is more effective and your ability to attract opportunity instead of distraction is perfected.

The ability to manifest is grounded in the way you live your life. It is not a recipe to grant wishes but a gift that all of us have, as long as we live a life with good vibrations.

Hard work and effort are aspects of resistance – you are overcoming, sacrificing, or struggling against something and it does not feel good. Life should feel good to you! You can achieve all your desires without struggle, hard work, and sacrifice, all you need to do is change your mindset from resistance to allowing. Allow good things to come to you easily and effortlessly, know that you deserve them, go with the flow of joy and well-being, allow it to take you to where you want to be!

Changing the mindset can be challenging initially like any other skill, but the more you practice the easier it becomes until eventually it will become natural to you and run on autopilot. As you rewrite the neural pathways in your brain to those of abundance and prosperity, you will begin to think and feel more and more in those terms and the thoughts of lack and scarcity will begin to dissipate.

People often ask how they can change their mindset and start thinking and feeling abundant when they are stuck in a job, they hate their low pay, when they are in debt and struggling to pay their bills every month when they only have to look at their bank account and see how "non-abundant" it seems to be.

What are Vibrations?

We all give off vibrations, colloquially called "vibes." You may meet someone and think, "They give me good vibes." But what exactly are these vibrations? They are energy. Everything in the universe is made up of energy at the subatomic level. The types of energy we emit are based on our thoughts and the types of energy we have within ourselves.

Normally, we do not notice or pay much attention to this energy. But this energy is not only a vital force it is also with us throughout our lives. Certain emotions can instill vibrations and thought patterns that we don't want, while others can instill vibrations and thought patterns that we do desire. If your vibe with someone, chances are you like how they think, how they see themselves, and how they choose to look at the world.

Identifying the Contrast

Most of us know what we don't want but aren't clear about what we do want. The distance between the two is in contrast. It's imperative to be clear about what we do want so that we can focus our energies and visions on that to attract it. How do we begin this process?

Begin to notice what types of thoughts and activities arouse positive or negative emotions in you. Then note the contrast in your emotions and physical state when you think about these different experiences. This type of awareness around your thoughts is central to clarify what you do and what you do not want in your life.

Love and Fear

Ultimately, all emotions can be reduced to either love or fear. Love emotions are the ones that create an open, relaxed feeling in the body: joy, happiness, contentment, pleasure, gratitude, hope. Fear-based emotions are the ones that produce feelings of tightness, pain, and closure in the body: anger, resentment, jealousy, fear, sadness, guilt. The more aware you become of the thoughts and actions that produce love, the easier it will be to invest in these ideas and experiences.

Ancient wise philosophers believed that everything moved, and nothing stopped, which has now been proven by science. All particles big or small spin and move, therefore they vibrate, all life is a great vibration that changes its frequency depending on the type of matter it is.

The LOA tells us that since all things vibrate at specific frequencies, similar frequencies attract each other just as the saying goes "like attracts like". The vibration of happiness attracts more of it in your life and the vibration of sadness attracts more sadness and stressful situations into your life. The ability of the vibration to attract similar energy is its resonance.

The universe does not judge your energy instead gives you more of what you emit or vibrate. It is like ordering in a restaurant. To understand what you are asking for, look at the state of your physical life. Relationships, money, vacations? You can always change your order, by consistently choosing different feeling thoughts.

By thinking of the energetics of a desire, you tap into its vibration. Then you practice the feeling of that vibration, therefore you begin to resonate with it. What do you think love feels like or what do you think a million dollars feel like?

Take some time every day to practice how to bring that feeling to life in your mind. Remember that thought and visualization create the feeling, the feeling creates the vibration or energy of it. Practice that consistently, tap into the feeling, become that vibration to tune into it. Like finding a station on a radio.

The frequencies are all out there. The million-dollar station, the love station, the better job station, the vacation station. All frequencies always exist; you must practice feeling into them daily. Becoming conscious of the feeling, identifying your mind with the feeling of the energy you desire. The thought is the order, the cooking is the practice of the vibration, when the food is done your order will resonate with you and be served to you.

The biggest misunderstanding of the LOA is that people assume that if it's true it should work quickly. There is always an incubation period, ideas must be first decided upon, then watered and nurtured so they grow (imagine their regular fulfillment). Like a fetus, the idea will manifest when it's ready to be born.

Since everything is energy this means that everything is connected to all things in the universe. The clouds, the water that fills the oceans, the animals, trees, me, you, and everything comes from one source and everything will return to that source. Feelings and thoughts are energy, too. Whatever you feel and think will influence everyone and everything that lives on this planet. If this is true, then we can create our realities since our minds rule over matter.

Quantum physics has offered evidence about this basic fact. It is a huge concept with extremely massive implications. It will give you a migraine if you try to understand all of it. It is not anything that most people even think about. It does explain our basic existence if you can mix the spiritual aspect with it.

This information is helpful since it is the foundation for figuring out ourselves and the world. This can open us up to life's bigger picture. The hard part is finding a way to explain all of this without completely blowing your mind while making this knowledge useful so you can get some benefits out of it.

When trying to understand all of this you will hear a lot of words like resonate and vibrations. This is just an easier way to explain everything. It tells us the ways our senses interpret all that is around us and translates it through vibrations. Quantum physics explains this as "an invisible moving force that can influence our physical realm."

It tells us how we live with the illusion that we limit our world because we think that everything in the world is what our senses can interpret. Once you can understand that everything is energy, you have to learn how to relate that to your personal growth and the law of attraction.

Our five senses cannot fully perceive the fact that everything is energy. Once you fully understand that everything is energy, you will be able to recognize it as an illusion.

Although our senses are telling us this cannot be true, it is. There is only energy, some people refer to it as source energy but it is there, and it has various changing forms and frequencies. The main difference is the way energies interact with each other.

Energy will interact with everything at all times and you will never see it. You might be able to feel this energy. Everyone is familiar with talking about our emotions or physical energy as being "energy in motion."

Humans have a wonderful gift of being able to control our thoughts. Everyone has free will and can choose where and how to direct their energy. This is a very small percentage if you compare it to the amount of energy we use in unconscious thoughts.

Our thoughts produce an extremely specific vibration and then this energy will try to find a match to that vibration. You will not even be aware you are doing it. This is similar to pinging a tuning fork. It can cause an object that has the same frequency if it is near enough to vibrate and they will be in vibrational harmony.

Every single object in the world has its unique vibration and all the energy in the world vibrates. A basic law of attraction states energy will attract itself to other energy that it resonates with.

Both the physical and nonphysical features of the universe are nothing but intelligence and vibrating energy. Nothing ever rests. The main difference between the things we can perceive whether they can be seen or not is how fast it vibrates.

Have you noticed how things seem to happen in waves? You are sitting at home thinking about something when a wave of emotion takes over your mind and body. When you open your oven door, a wave of heat hits you in the face. If you listen to a thunderstorm, you can hear a sound wave of thunder. If you are outside on a windy day, you feel the wave of air as the wind blows your hair. You are outside working in the garden and you can feel the waves of the sun as they fall on your skin. You are walking along the shore when waves of the ocean roll over your feet. All kinds of energy will travel in waves.

This energy can't ever be destroyed or created; it only changes forms.

Thoughts are also energy. The brain is the most powerful electromagnetic tool that was ever created. Don't worry, you don't have to be a physicist to understand this but when you can begin to understand the concept of everything is energy, you will realize you can break this down into usable levels in your own life.

Since humans have free will and consciousness, there are a lot more layers involved since we can create our beliefs based on the things that we were taught growing up and what we have experienced in life.

This is our power it can be a challenge since we have a limiting belief system that obstructs the flow of energy inside us that we aren't even aware of most of the time. Our belief systems are also energy, and this is what gets in our way when we try to attract things to us.

Our belief system is limited to the energy we put behind it and this means it can't be changed. When you finally realize that your belief system is not working for example: "I will not ever find a job." Or "I will never be able to get out of debt.", you could use therapy such as EFT to get rid of certain problems so you can be open to new realities and possibilities.

You are probably going to find some areas of resistance that you don't have to label wrong or right.

It only appeared because of your point of view or beliefs at that moment in time. Think of it as a gift if you find them. Notice it and be grateful you had an opportunity to change it into something that you liked.

Everyone is vibrating some kind of energy, and we will attract it into our lives. We might think that we are focusing on the things we want but beneath it all; we are vibrating, believing, and focusing on the things we don't have. If you can get rid of our low vibrations and subconscious blocks, it will allow you to choose new actions and expand the mind that will then create new results.

Focus on your life and see everything you have: your health, your house, your bank account, your job, your partner, and your children. These are all things that you created in your life. Now go back and try to remember the main beliefs and thoughts and see if these match up. If you are completely honest, you should see a correlation between how you thought your life would be and what you thought was possible or what you thought you deserved and were capable of achieving… you might see that your life reflects these limits.

Now that you understand everything is energy, and you realize that you are nothing but energy, you might feel empowered about your future transformation. If you still don't completely understand, give it some time.

But is everything really energy? Is everything in the world connected? If you looked around at this moment, you could see your television, your laptop, a window, a pet; outside you might see some trees, flowers, or your car. Yes, all of these are individual things. Since everything contains the same matter, does this make everything one? The truth is what I see is many phenomena instead of a unified whole. You might have learned that all matter is made from molecules during your school years. You probably have learned that all molecules are made up of atoms and that atoms contain a nucleus that has electrons orbiting around them. All of these are particles, fractions, and parts. But when you first see it there is not a unity, a whole. It doesn't matter because appearances can be deceiving. What might look like a solid may not be as solid as it looks.

Scientists have proven that matter has 99.999999999999 percent empty space. Yes, there are 12 nines after the decimal. If you were to make the nucleus of an atom as large as a pinhead, the first electron would be 160 feet away. You will find nothing but empty space in between. This means that what you are reading right now, whatever you are sitting on, your house, the Earth, all this "solid reality" is mostly empty spaces. That's a lot of empty spaces. So, what exactly is left when talking about solid matter? A calculation will show you that the solid part of an atom is just 0.000000000001 percent of the entire atom. It is hard to believe that all solid objects would consist of this little bit of solid matter. If you look at it that way, anything solid really is not solid.

No Solid Parts

After scientists discovered atoms and molecules during the 1600s, they just assumed that atoms and molecules were made up of solid particles. If you look at quantum physics that was developed during the early 1900s, there are no "particles." Energy is the backbone of all reality. Each particle is considered a vibration. Electrons will vibrate in an electron field. Protons will vibrate in the proton field. Since everything is energy, this means that all things are connected to everything else. On an elementary level, the matter will not show up as particles that are isolated. All matter is a dynamic connected tissue that is vibrating energy fields. Solid matter and atoms are made up of mostly empty space; it's the same out in deep space. There is the same amount of sand on all the beaches as there are stars in the Universe. These are infinitely large numbers but in between these are all that empty space. All that unused space is wasteful. Scientists in quantum theory found that particles contain energy as well as the space that is in between everything. This is called zero-point energy. This makes the statement: "Everything consists of energy."

Zero Point Energy

Scientists have discovered there is more than just primary energy that works in the whole Universe and connects all things to all things out there. This energy was first measured by Dr. Harold Puthoff. This experiment was done at zero degrees Kelvin which is known as absolute zero. If you boil anything by adding energy to it, the molecules begin to move faster and faster. For example, when you heat water, it begins to boil and then evaporates. The opposite of this happens when you freeze water. The molecules begin to move slower and then the water will become solid when it freezes. According to an old scientific method, no elementary particle, atom, or molecules can move at absolute zero. This means at this temperature it shouldn't be possible to measure energy at all. However, instead of finding no energy, he found an abundance of energy that he called "a boiling witch's cauldron."

Energy That Has a Thousand Names

The science community has just recently figured out what the ancient culture has known for thousands of years and that is that energy can penetrate matter. All things come from energy and then they return to that same energy. This is the source of all life forms. Every culture gave it a name, and this is where it got the name: "The energy with a thousand names." You might be wondering what this is. Well simply, it is life energy. Life energy has been known to every culture since ancient times and each culture has a special name for it. The Chinese call this *Chi*. The Greeks call it *pneuma*. It is known as *spiritus vitalis* in Latin. It is called prana in yoga. It is known as *mana* to the Kahunas in Hawaii. It is the Christian's Light. It is the *ki* of Reiki and the *ka* of the ancient Egyptians. In today's culture you will encounter this energy as the *orgon* of Wilhelm Reich, the *od* of Reichenback, the *aether* of anthroposophy, and the fluids of mesmerists. In each case, they are talking about universal, subtle energy that can penetrate and included all things. This energy is what carries life and connects everything. It forms matter. Life energy is normally mentioned with life force. It has magnetic and healing properties.

Information and Energy

All energy fields contain information. You can compare this to scripts that actors use to keep themselves organized. This field informs matter and becomes matter. It is similar to incarnation, where a person's soul manifests itself into another form. Information does the same thing. It turns into a form. To read a book, you have to digest the information and then act on what you have read; you are going to need energy.

Many people treat information and energy as the same this generates a lot of information. You have to realize that information and energy are two different properties. A newscaster gives us information. If we have problems hearing them, we turn the volume up. At this moment, you are adding energy; this will not change the news or the information. You need to exert energy to get the information from the speaker out to the audience but the information will stay the same. Nobody would think that you are telling a different story if you started speaking louder. You can buy a book about anything but if you don't read the book, you will not get any information from it. Once you have read it, you will need to use energy and time to act on the things you have read.

Vibrations of the Universe

Three things go on inside you which are commonly misunderstood. The first is the thoughts you have; the second is the inspirations you receive, and the third is the feelings you get. Your thoughts are generated by your brain/mind. Your inspirations are the vibrations of the universe that you feel deep inside but are relatively silent. And the feelings you get are just primal instincts that the body generates. You need all three working well and interpreted correctly because the best gifts you get from the universe come to you in the form of feelings. But we usually get them wrong because we confuse them with the feelings generated by the louder body.

Distinguishing Learning from Inspiration
Do you meditate? If not, then it's time to reevaluate your entire life strategy and your pursuit of success. As it is, if you are not meditating, then it is very likely that you are missing many of the parts that will get you the superior result for the correspondingly appropriate level of effort. It is also highly likely that if you are not meditating, you are constantly hitting a brick wall of failures and you have to run two lives - one that is focused on the goal and the other where you are constantly cleaning up after failures.

There are two ways to succeed in life. Success can be yours either by allowing the consequences of failure to point you in the right direction, or you can get the answers you need to get on the right track and achieve what you need without fighting the consequences of your failures.

Don't get me wrong. Failure is not bad, it is the greatest teacher there is. It is there to knock down for you the secrets of the universe, so that when you aim for something, you end up getting it.

Feel the Universe in Meditation
But meditation takes a different approach. Meditation prepares you to receive the whispers of the universe around you so that when you make a decision, it is the correct one. You just know, almost instinctively, what to do in every particular situation.

The single most powerful practice a person can do is meditation. If you subscribe to the school of thought that meditation is about calm, Zen, and peace, or prayer, devotion, and spirituality, then there may be a reason for some of you to walk away. But meditating is neither religious nor spiritual. It can be if you wish it by itself, it is what it is.

One way to view meditation is to consider it as a force that brings harmony to competing systems of the body and mind as mentioned in earlier chapters. The will of the body always seems to go in one direction, while the will of the mind always seems to go entirely in the opposite one. Many times we give in to one or the other and feel a sense of remorse that is nothing short of distracting. But the debate between the two can be chaotic and distracting. One of the effects of meditation is that it aligns your competing systems and brings harmony to the chaotic voices. That removes distraction and brings peace to the whole.

Have you ever had the chaos of a raging debate between reaching for a cigarette twelve hours after quitting? The battle that goes on is between your 'body' that craves the nicotine and your mind that decided to quit cold turkey, or the debate of wanting to buy the latest sports car. One side of you wants to put you behind the wheel, while the other side is telling you that the right thing to do would be to save the money and instead get something you can afford.

Each person has been in numerous pairs of opposing forces in their life. Sometimes we give in to one force, other times we prevail with the other. Whichever decision we make, the future that arises from that decision is colored with the consequences of that action. Not taking that cigarette puts you on the path to one set of consequences, breaking your promise to quit, puts you on another path. We have all experienced that.

If you look closely at these battles and chaotic arguments, you will begin to see that there is a third actor on that field of battle - a silent umpire of sorts. Some people call it a conscience, others call it the soul. But those words carry with its significant baggage that distracts from this discussion of meditation and success. For now, let's refer to it as our core. However, you first need to identify what we label as the body, and what we label as the mind.

Much of our perception of the world comes from our senses. We see, hear, smell, and so on. On its own, none of that means anything. On its own, our senses are just sending electrochemical signals to our brain and it is our brain that translates what it is and what it means. Whether it is sight, sound, or smell, if we have no prior introduction, we will have no idea what we are looking at or what we are smelling. We sense non-binary data through association.

Binary data is the type of data that can only be one or the opposite. Yes or no, on or off, right or wrong. There are no shades of gray and in-betweens. In the beginning, the central nervous system could only be in one of these binary processes. But we evolved, and a few million years later, we could conjure more complex thoughts and thinking patterns. That thinking pattern even allowed us to be able to produce future events based on present factors. Our brain becomes more powerful as it developed the mind and the mindsets that were built on top of that. We could even turn the observation on to ourselves and learn who we are and what we are about, we could finally turn consciousness back onto itself and ask the questions that elevated our existence. This higher level of thought is nothing short of amazing.

But at the back of all that, right down there at the base of the skull, sits the ancient part of our brain, still calling some of the shots when certain circumstances present themselves. One of those hard-wired events that result in fear is when we face the unknown. Fear is a powerful primary emotion that developed to stay alert from their predators. That still sits with us today and can disrupt our entire thought process and the outcomes that follow.

However, if you can learn how to tap into your soul (again, there is no religious reference attached to the meaning of the soul) then you can overcome the crippling nature of fear.

As we embark on a journey across this book, you need to understand its final purpose. This book is about making you all that you possibly can be. This book is about putting power back into your hands. It is about showing you the source of limitless potential and understanding - it is about reaching an awakening. It is about showing you that there is a teacher within you that you can access. If you can look there and ask, whatever you ask for will come your way.

Meditation is not about lighting candles and closing your eyes. That is not meditation. Silence is not powerless; in fact, silence is one of the most powerful resources. If you can tap into that, all you ask for, the greatness you envision, the power that you need are yours for the taking.

Three steps make the path to success. It is this path that you must learn and internalize if you are to truly escalate your abilities and elevate your life from here to greatness in one lifetime.

The first step is the development and execution of mindfulness. The second is the development of the discipline required to focus. The final step is the various levels of meditation.

Once you can do these steps effortlessly, there is nothing earthly that you cannot accomplish when you set out to do it.

Fourth Dimension

For a normal human being, realities are limited to past, present, and future time modules. He reads its surface-conscious life, he is limited in its capabilities because he cannot see the whole picture, he does not stand atop a mountain but he is still a mountaineer. Past and future are something of the imagination of the physical mind for the higher self, all-time modules exist in a framework of a complete present. He can back off and introspect, foreshadowing tomorrow.

The fourth dimension is the integration of past, present, and future within a frame. The relationship with the higher self opens up the fourth dimension because until you connect with the higher self, your physical existence will hold you with different views of past, present, and future.

We live in a three-dimensional physical world and the fourth dimension sits within us. It is not in physical form but it is concerned with the soul and the subconscious part of our being. To feel the presence of the fourth dimension it is necessary to link between conscious and subconscious parts. In simple words, it is the link between the brain's thinking (brain) and the soul. The fourth dimension strengthens the influence of inspiration, knowledge, spirituality, and creativity in one's life.

A mindful connection between the brain and the soul connects the conscious and the subconscious which eventually takes the fourth dimension to a conscious level where you can feel it. The fourth dimension is with conscience and glorifies the importance of welfare, harmony, righteousness, and positivity. Fourth-dimensional thinking brings you closer to your divine self, enlightens you and helps you to manifest your greatness.

Those who manifest their greatest form appear from a higher state of consciousness, from which they receive inspiration or creative ideas from your higher part to bring something into their life. It often appears as an inspiration or a creative idea. When it comes from their higher part, inspiration carries with them all those who need to disclose energy, vision, clarity, enthusiasm, people, opportunities, resources, and guidance. This is why fourth-dimensional thinking is fundamental to expression. Manifesting from a higher level of consciousness allows everything in your life to manifest according to your deepest being. The shifts that come from expanding one's consciousness and knowing one's relationship as a part of the entirety of life will manifest in many ways.

Tips to awaken the fourth dimension

Everything that ever exists in the fourth dimension is beyond our physical existence. Here are some tips that you can develop by adopting a quadrangular mindset:

- Become an observer and observe every possibility that may exist. See the bigger picture in focus.

- Learn to casually let go of the need for something, believing that you need what is already there.

- When you want, ask for your best. Rise above the materialistic thought process and seek the inner light, seek the meaning of life and divinity.

- Always pay attention to the core. Your soul always wants experience while your physical mind wants something material. Pay attention to the essence of what the item gives you.

As the light of consciousness begins to enter, the refined third-dimensional thinking seems limited and meaningless. Anyone can become angry or outraged at the society that has spread the word. In the fourth dimension, vice versa and the judgments still occur but much less physically. The subject is spirituality in every context. Awakened conscience is necessary for this.

Vibration Profile

Your entire being, from the tissue of your beating heart to the tips of your fingernails carry a certain vibration because they are made of matter, and we all know that matter is the vibrational state of energy. Energy can be recomposed to be force and matter. Force can move matter, and matter can carry force, which you know to be momentum. Think of ice floating in the water. The water can move the ice, and both are the same thing at different states of vibration.

Underneath space and time is a fabric that connects every animate and inanimate object, from people to thoughts, from planets to empty space. Science has finally confirmed its existence, something ancient meditation gurus have known all along. In scientific parlance, they are called fields, and each fundamental particle has its field. A vibration in that field gives rise to the building blocks of matter. When they test the Large Hadron Collider in Switzerland it is not to find particles, but rather to find the fields that give rise to these particles. And they have been immensely successful.

By now it must be apparent to you that you are indeed an object that is in a constant state of vibration. Like a tuning fork. Through this field that's vibrating to make you, you are connected to everything else in this universe that is made of energy and matter. You are as connected to your twin as you are to the rock salt buried deep in the Himalayas.

This connection to everything and the vibrational profile that makes you, and everything else what it is everywhere, is how the Law of Attraction works.

Your vibration is the basis of everything that you do and who you become as you take your steps from cradle to grave. But it is not fully autonomous. It needs input from you to give it direction. Imagine a car going down a hill. It will keep going, regardless of your opinion. It will get to the bottom of the slope. If you take control of the car you can dictate where it will go. In life, you will attract all sorts of things into your life without you realizing it. The thing you have to do is control where you get the results of this power.

You vibrate at a given frequency no matter what you do. But you can control that through various means. Exercise is one way to control your vibration. Meditation is another. Exercise, dance, techno music are all ways to increase your vibrational profile, while meditation, sleep, classical music (think Bach, Vivaldi, and Tessarini) will lower your vibration. Walking is also a method of manipulating your vibration profile. You have a vibration profile that you are born with, that is the vibration that manifests as you, but you can move that around in any number of ways as described. Each individual has to find the way that bests control their vibration. A good rule of thumb is to get the appropriate activity for the individual to control his vibration is to look at body type.

Determining Your Vibration Profile

There are three general vibrational body types. The first is the person who can never seem to gain weight no matter what he eats. These people are highly strung and very energetic. If this is you, the best way to control your vibration is to learn the practice mindful meditation. You want to tune it down.

If you are the chubby type and no matter what you do you can't seem to lose weight and have very oily skin, then the best way to affect your vibrational energy is to do aggressive workouts at least once a day followed by deep breathing exercises. You want to raise your vibrational profile.

For the third body type, you are the one that is neither of the first two. To get your vibration in order the best exercise for you is to eat frequently, whenever you are hungry, get lots of fresh air and meditation exercise will bring your vibration into the zone you are looking for and get you to where you need to be to control that vibration. Stay away from caffeine. For this body type, you need to be able to control the swings of your vibration from getting too high or too low.

Regardless of which body type you are, meditation will always help. A combination of reflection, mindfulness, and meditation is always beneficial in getting you to the right vibrational frequency.

The different regimens to achieve optimal vibration are based on simple reasoning. Those with different body types operate at different vibrational levels. The higher the vibration, the more active the person's persona, and correspondingly they tend to have higher metabolisms and don't seem to gain weight easily. Low energy types have lower metabolisms and thus are a little heavier.

Once you figure out your body type, you can set about your task of finding your optimal vibration by varying the frequency of your workouts, and by balancing it with meditation. If you are into yoga, that will work to your advantage.

The reason meditation and mindfulness also work for all body types is that you can have meditation strategies that increase your vibration. The optimal vibration is reached when you feel at peace. Meditation automatically brings about the state of peace by raising or lowering energies without the practitioner attempting to figure out if they need to raise or lower their vibration.

Remember, it's all about the vibration and once you get into the habit of striking the right frequency of vibration, your ability to invoke the leading edge of the Law of Attraction will come as a natural progression to your efforts.

Attuning to the Vibration of the Universe

Attuning to the vibration of the universe is a necessity to get the LOA to work for you. It increases your physical charge, which amplifies what you put out mentally and verbally. Raising your vibration is all about how you feel and making yourself feel better. The point of life is to be happy therefore, you must take time regularly to make yourself happy. There are a few different ways of looking at this.

When you're in a good mood the vibration you're emitting is already high and in general that's where you always want to be. Being in a good mood is pretty much the ideal place to be in life. Being in a good mood usually reflects in your environment where everything just seems to go right and you're going with the flow. The only problem with that is it's hard to always be in a good mood, so we must work at it.

Getting in a good mood is raising your vibration. You probably already know what helps you get you in a good mood. Listening to good music, exercising, meditating, going for a walk etc. Although some ways are better than others for optimizing the use of the LOA. Dancing is hands down the best way to raise your vibration.

Good music coupled with physical movement allows your body to release tension and your mind to become one with your body. When the mind and body become one due to physical activity you can enter a thoughtless and emotionally rich state of mind. Just like a runner's high.

Take advantage of when you are driving and listen to music you love, music that makes your blood flow. Dancing also happens to be one of those activities you might do more if you were already living your "end" result. Using this state of being is the best time to use affirmations mentally or better yet out loud.

Just think of Native Americans dancing and chanting around a fire. They are creating a charge of energy to increase their energy to tap more easily into what they are trying to accomplish. Another example is the whirling dervishes in Turkey, this is a specific dancing style that puts its practitioner in a heightened state of awareness that can create a very happy state of being, or in this case closer to God.

If you practice martial arts you may notice that some styles are based on a yin and yang concept, body movements that flow perfectly into one another in a very rhythmic manner, just as if you were dancing. Traditional forms of martial arts are based on the natural motions of the body. That's why some martial art masters can move extremely fast and be so strong with small motions like Bruce Lee's one-inch punch. He understood body mechanics perfectly and mentally became one with its flow.

Points to Take Away

Goal setting is key to reaching a goal, it's a literal target. Take time and consider all the details necessary to create a fulfilling and possible goal. If you are just starting with the LOA, you need a goal that is somewhat realistic for comfort level. When the target is reached, you can always aim higher. Having a major goal is what you visualize, meditate, and focus on daily.

The minor goals come from the major goal and become the steps to take to make the overall goal a reality. Do what you can every day that will lead you to "the end "result, take those small steps.

Daily meditation or taking the time daily to focus on the end result is key, losing sight of your goal means you don't care enough about achieving it. If that's the case, it's ok but you need to find a goal that excites you, a goal that can fulfill a deeper meaning in you.

Just like an Olympic athlete you need to do something daily for the goal to become a reality. It is worth taking time to do a 15-minute daily meditation. If that is too much time for you; you are not recognizing the importance of focus and or of the power your subconscious mind holds over your conscious awareness. Taking time to focus on a target is like recalibrating a weapon that helps guide you towards the target, in this case your brain.

Spending time with yourself helps you acknowledge your true desires out of life. Time alone brings light to what is to be human and it makes you more compassionate and relatable to other people. We all have similar minds, to acknowledge how yours works can give insight into others' minds.

Finding ways to specifically raise your vibration while being alone can lead to good things. Exercising, taking long walks, listening to good music, and dancing all let a deeper wisdom leak through to your conscious awareness. Raising your vibration while practicing visualizations or affirmations also raises the vibration of those ideas.

Making a habit to feel and think differently is key to creating a new life. All that now surrounds you is from past thoughts and beliefs. To create a new future, you must regularly be and think about the change you want now regardless of the environment that surrounds you. Be visionary, be the change.

Exercise 1:

Make a list of five things that you can do to help your business grow. Use the following questions as a place to start.

Can you join networks?

Can you reach out to colleagues and see what they did to grow their business?

Do you have a website?

Should you update your website?

Do you need to update your office?

Can you learn a new skill?

Do you need to hire someone?

Do you know how to do that?

Make a list and follow through with all of them.

Exercise 2:

Practice each style of meditation and stick with the one that feels right to you, spend time daily doing one or all of them. Morning and bedtime are the best time to visualize your "end" goal and saying affirmations.

Sometime after fully waking up is a good time to watch your mind. You may need more mental energy to learn to detach yourself from your thoughts.

Exercise 3:

Take five mins here and there throughout your days to refocus yourself on your vision.

Learn to step away from negative environments that take over your mind, use your emotions as guidance to understand these times.

Exercise 4:

Spend some of your free time doing the things you love to do, try not to put them off, for life is shorter than you think. Try to do activities that make you feel like you are already living the "end" result and when you do them try to feel gratitude or appreciation for it.

Mindfulness

There is a lot of literature on the practice and discipline of mindfulness. And just like there is the unfortunate misunderstanding of what meditation is, there is, even more, a misunderstanding of what mindfulness is.

You need to know three things to be able to use mindfulness as a step to meditation. The first is that mindfulness is about organizing space and time. The second is that mindfulness is about the alignment of your conscious mind with your current space and time. Finally, mindfulness is about expanding your conscious ability, which begins weak, and if built up can enlarge to great power.

There are two cosmic phenomena that we will introduce you to so that you can understand the scope of your mindfulness and how it ties in with the entire universe.

Space and time are two measures of an environment that spread across the universe. Beyond the boundary of the universe the dimensions of space and time cease to exist. The energy that defines the expanse of this space and time is what gives rise to matter and all that comes from the existence of matter.

Space is beyond what your ruler can measure, and time is more than what your clock measures. We know that Einstein theorized, and physicists have proven since then, that time is an individual phenomenon and is not standardized across space.

It would not make too much sense to delve too deeply into the physics and nuances of time, except to say that we need to look at time as a stream. The current analogy works in many ways but it is not one that you should get carried away with. It has a purpose here beyond this, it can be counterproductive.

The reason we look at time as a stream here is that time flows in one direction, and that is important in the quest to remain mindful. Second, time is as wide as a river, or as narrow as a stream, and the width of time that you can see depends on how expansive your conscious mind can be. The final reason the river analogy works is that it allows you to see things that happen across space as a passage of time - just as the passage of the river.

Mindfulness from Space and Time

There are millions of things that are going on around us. Imagine if you could have a screen looking at everything going on in the universe today - from the hatching of a fry to the birth of the sun in a distant sector of the universe. That's a large number of data streams - a very wide river.

On the other hand, imagine if you just have one screen and that on a screen it shows you breathing. Nothing else, and as far as you can see, the entire universe, to you, is reduced to one act - the movement of air in and out of you. The first resembles an infinitely wide river with large amounts of data; the second represents a very narrow river. If our capacity is limited to be able to understand and have a handle of just one stream but you are given a mighty river to handle, your mind crashes. You become overwhelmed and you can't fathom your surroundings.

The problem with being overwhelmed is that the moment you do, the circuit breakers in your head break and you cannot process your surrounding and whatever is facing you. We know that feeling all too well. The moment you cannot understand your environment, a primitive emotion takes over and that is the emotion of fear. People are built to naturally fear things they do not understand. When you are overwhelmed, you will end up in a state of fear, or in some people that can even lead to anxiety and panic attacks.

By invoking mindfulness, you can slow down the flow of data and ease the feeling of being overwhelmed. Mindfulness is not just about focusing on one thing at a time, as most people will tell you online. Mindfulness is to focus exactly on what your breadth of the stream allows. Most people start with just one thing at a time. You can do the same, then once you have practiced it for some time, you will find that you can indeed multitask - just doing it one thing at a time. The key is that your thoughts and your actions are one.

The crucial thing about all this is that the thing you focus on should be in the present. It should never be about the past, and never about the future. And since you can only be in one place at one time, you will automatically focus on the place that you are in, at the time you are in. When you do that, the entire breadth of your stream is applied to that moment.

When you apply your mindfulness to one thing, and not forward or backward in time, what you can do is to look at the one thing more deeply than just the superficial characteristics of the matter. As you develop your mindfulness, you will find that you can penetrate the subject of your attention deeper and gain a visceral understanding.

The more you practice that, the more you will strengthen, or in this case, expand that stream. When you widen that stream, you can apply more of you to the moment than before. This will continue to grow, and you will continue to see things you could not see before.

The size of your stream compared to someone else's will always be different. Some people can handle more in their conscious mind in one moment than another person in that same moment. When you practice mindfulness, you start to expand that.

It is rather simple to be able to start with mindfulness. You can do this almost anywhere by doing anything. After all, practicing mindfulness is just doing and thinking at the same time and going deeper into its processes, rather than going wider. But if you insist on being more disciplined about it, then grab a comfortable chair and sit down. Close your eyes (this is to cut off visual distractions) and direct your attention to the air that enters and exits your nose. Your breath is a natural metronome that allows you to bring chaos back into rhythm. This is a time-honored practice, and you should feel very comfortable doing it.

Do not control the tempo of your breathing; instead just let your body do what it does best. You just have to sit back and watch it. If you have trouble watching it, then sit back and count it. One for in and the same for out. Keep counting to ten, and then start again. After you get the hang of it, you can stop the counting and just observe your breath.

Once you can do the short sessions, then it is time to expand your abilities. The next step is to take the sessions from your couch to real life. You can apply mindfulness to everyday tasks, from the time you wake up to the time you go to bed. If you can transplant the same practice from breathing to doing other things, you will find that you can do them more efficiently and more accurately.

At this point, mindfulness will come naturally to you and you will discover how much more you absorb at every waking moment of your life. You will also find that you do not even need to take notes in a meeting because you can remember them just by being mindful of those whom you listen to. You will also find that you can remember names and faces. Your capacity would have a marked increase.

Even once you can be mindful, you should still practice it daily on your own. Mindfulness is not meditation, even though it is used as a path to meditation. But it is a discipline that you use to sharpen other areas of your psyche and mind.

Exercises in Mindfulness

There are three exercises that you can use to develop and maintain your ability to be mindful. If you graduate mindful training and get to the point of successful meditation, there will still be days you will need the extra help to be able to meditate. A day will come where you will not be able to meditate or be focused. That is when you will find these exercises helpful.

Stage One

Breathing. We covered this before and it is usually enough to get you going. But here is another way you can use breathing to get more grounded on awareness.

You can be anywhere for this. It doesn't matter if you or on a train, or in a coffee shop. You can keep your eyes open or closed for this, that's up to you. First begin inhaling deeply, then pause followed by a controlled release. Do this three times and, each time you do it, pay full attention to the point of your nose that meets the bridge just beneath your forehead.

Imagine looking at that place. If your eyes are open, it would almost seem that you are cross-eyed.

As you look at that spot and complete three sets of deep inhalations, return to normal breathing with your eyes closed. Count each breath up to 10. Then start at zero again. Count again, up to 9. Then return to zero. Count again, up to 8. Then back to zero. If you lose track - start all over again, from 10. This should take you about five minutes but each person is different, and you shouldn't speed up or slow down to keep within this time. Relax and do it at your own pace.

When you finish, and you open your eyes, you will be amazed to find that your disposition changes, and eventually your position on things seems fresh. You will be tempted to take it all in but try to apply your fresh state to the things that require your attention.

Most successful entrepreneurs who do this get straight to work on areas that need their creative powers. Or they do it before a workout.

Stage Two

The next level that you can try is a lot less complicated but requires you to get an app for your smartphone, or you could simply set a recurring alarm for every minute. In this exercise, a bell will ring at specific intervals. Personally, my phone chimes gently, every twenty minutes. Twenty minutes is my magic interval and seems to work well for me. I am sure that will change as time passes.

Twenty minutes has been my thing because it is my natural window of efficiency. Whatever I get done usually only takes 20 minutes. And at this point, I am making sure that my focus is updated and renewed for one of two things. In some cases, it is usually time for me to move on to the next task. Or it's time to inhale and strengthen my mindfulness for the same task.

I have found this to be invaluable in getting my efficiency up. Many of us who practice this have found that there is no correlation between our effective window but one thing that has held to be true is that each person is different.

At this point, the alarm sounds. Close your eyes, take three deep cleansing breaths. Once it's done, get back to what you need to be doing.

Stage Three

This stage requires that you learn a new practice. This is the stage where you bring mindfulness to your entire life. Not just every evening or every hour. This is to get you to turn your time into a full nature where you are not in one place while your feet are elsewhere. You need to be in the same place your thoughts are. Planning is not considered going against this, nor is evaluating the future of something you are considering. These are all voluntary directions of the mind. The point is that you are not thinking about something other than what you are doing.

You do this by making sure that stage one and stage two are easily done. That takes time and practice. It takes diligence on your part to make sure that these things become habits. The ability to form a habit is useful but that habit is not about doing things on autopilot, the habit is about reminding you what you need to do at a particular time.

You start your day by reminding yourself that you will spend the entire day in a state of mindful focus (the focus part will come later). You tell yourself that this is your path to success, and it requires the least amount of effort but has the most amount of bang. From that point, you stay mindful of everything and you be ruthless about it. Your path to meditation and mindfulness is not about making you tame like a rabbit. It's meant to make you as devastating as a lion.

Nothing about this meditation and mindfulness is meant for you to mistakenly think that you are about to become meek and nice. It's not. You need to wake up that lion in you that will set your sights on one thing and get it. That's how successes are made.

Every time you catch yourself drifting, get yourself back to where you need to be. Use the breathing exercise to do it. If you find that you are distracted do a short workout to burn up the energy and get your circulation going then focus on that heavy breathing. As you feel the intensity grow, focus your calm on that. It's a form of energized calm that you will learn to appreciate once it starts working for you.

Strength from Within

Have you anytime understood that you, like some other individual, have a gold mine of characteristics where it counts inside yourself? By far most of us are clueless about how we have such hidden gifts.

Whatever is left of us, paying little respect to whether we understand that they are there inside us, don't have the faintest idea how to take the characteristics out away from any discernible block and put them to use to upgrade an amazing idea or in driving a more fulfilling background. Allow us to research ways and techniques to reveal our hidden characteristics to improve our lives.

In any case, you should believe that you have some novel characteristics.

Stop Negative Self Talk

Never say to yourself, "I can't." Instead, imagine yourself having inside you the needed qualities to fight any condition, paying little attention to what that is, and that you can, and you without a doubt you will, win if you do it properly.

Remember, self-conviction is assurance, and getting certainty is a large part of the battle.

The next stage is to start exploring yourself. Almost investigate your experience, and the genetic. Not that you have not done this already.

You may have. Little by little you will have to do it even more efficiently. Record what you got from your people or grandparents. If you don't think you have not obtained any; think about what they have told you.

Is it possible that in some place significant inside you, you have those excellent characteristics? In any case, you have not understood that you could use them. Summary all of the characteristics and capacities of your people or grandparents.

Check whether a couple or all of them can be put to use.

For instance, does music continue running in the family?

Did you ever observe that your mother had tremendous measures of steadiness?

You also may exemplify it, without observing it.

Have you anytime seen that you have, essentially, a way with words that others you know don't have? Have you not put this bent to extraordinary use? Maybe you are contributed with a strong body.

Be that as it may, you don't think about the habits by which your physical quality can be utilized beneficially. Research, attempt and, finally, abuse. That should be your framework in drawing out your characteristics from any restricting impact.

Concentrate on Your Qualities

Your acquired establishment seems to you to be progressively direct to analyze. Regardless, you need a more in-depth evaluation. Once again, prepare an unequivocal summary of the characteristics your guidance and planning have given you.

What is the scope of capacities you have?

Is it genuine that you are using your abilities and qualities to the fullest?

Have you looked for after your interests, changed over some of them into side interests, and thought of the probability of changing no short of what one of the later into a subsequent vocation?

List your inner qualities such as restraint, compassion, bargain, watchfulness, constancy, or confirmation, etc. to accomplish achievement in your life.

This is anything but a one-of-a-kind exercise. As your life keeps going, periodical reviews of your characteristics will allow you to pinpoint imperceptible properties that may help lead you down the right way. Who knows, you may strike gold!

Changed Observations

When you are busy structuring, enhance your character and traits, choose your verbal and non-verbal reactions to the events of this world. As an underlying advance, let us appreciate that your impression of reality is based on your subjectivity.

So also, as a guide is only a less than normal description of an area, what you see as true is only a hidden description of this present reality, not just reality. You can't fight the temptation to look at the world through rose-tinted glasses. Your reactions are overseen not by the reality yet rather by your view of that reality.

NLP (Neuro-Linguistic Programming) causes you in getting this and reducing, if not emptying, your subjectivity. You will at that point perhaps think about getting elective viewpoints of this present reality and, therefore, accomplishing a move in the way where you react to it.

For what reason do people react another route to a particular event or condition? Is it not because of the qualifications in their perspective on that event or occasion? What a terrible setback is to one may not be the same as another.

For instance, a couple of people may be upset by verbal or physical misuse or ignore it. Others may be so affected by it that they may require mental treatment.

The fundamental thinking of NLP is based on the principle that it is possible to change one's recognition, feelings, and lead to make frightful experiences justifiable. You may even end up impervious to damage.

The most common obstacle in manifesting our desires is that we tend to look at our current circumstances and derive our feelings from what we can see. For example, if we desire abundance, it is usually because we have experienced the opposite. We say we want abundance but then we look at the state of our finances or bank account balance, feel frustrated and focus on the lack of abundance. We become depressed and worried about what we see. Doing this automatically puts us in the mindset of lack (the opposite of abundance) and we unknowingly attract and manifest more circumstances of being broke. We then look at those circumstances and derive from them more unpleasant feelings about being broke. And the cycle goes on.

To break this cycle, it is important to understand that we mustn't focus so much on what it is but more on that we want. If we want abundance, we have to find ways to think and feel abundant now, regardless of what we are now seeing. This is where the concept of money is not always helpful. Money has a lot of negative feelings and beliefs attached to it, some of which we assumed early in our childhood. When our parents tell us "We cannot afford that.", "Money doesn't grow on trees" or "You were not born rich like so-and-so", we then assume and internalize these beliefs and they live hidden deep in our subconscious for years to come. We then often unconsciously believe that we don't deserve financial abundance, or that we are not good enough to have it, or that we are not lucky enough or young enough or smart enough to have it. Bear in mind that this is mostly subconscious, so while consciously you will say 'Of course I know I deserve it!', your subconscious has a different belief which is in contradiction to that, and it is the subconscious belief that it is more powerful and will sabotage your conscious efforts. It is, therefore, important to reprogram the subconscious mind but also it can be beneficial to entirely drop the subject of money.

If you say your goal is to be a millionaire, it is an exciting goal but you may find it difficult to believe it consciously if you are currently nowhere near that state, or never have been. Also how do you visualize such a goal? It can be somewhat limiting to only visualize bank account balances or heaps of money.

Instead, get off the subject of money and go more general. Consider instead of wishing to become a millionaire, wish for abundance and freedom. Rather than amounts of money, visualize all the things you would do with the money. Where would you live? What would your house(s) look like? What would you buy? Where would you go? Whom would you help? Get immersed in those ideas and the feelings of them. How would it make you feel? Relieved? Free? Appreciating? Grateful? Abundant? These are all the vibrations you are looking for to be at, to manifest those desires. Your mind doesn't know the difference between what is real and what is imagined, so as you think the wealthy thoughts and feel the wealthy feelings, you are literally reprogramming your mind to a wealthy state of being.

Manifesting

You can manifest anything, meaning you can take an idea or thought that you like and make it real in life. Wishes do come true if and when you have managed to clear all the blockages and stumbling blocks and get in sync with the universe. The Law of Manifestation that you need to know about is all around you if you observe the things that happen to you and around you. You know it is true; I am just going to show you how to make it happen. You still need to practice. Manifesting something is to receive that thing in this tangible world. It is a gift. No matter how hard and smart you work for it, it is a blessing, and you should be grateful for it. Once you are, you will get more. Once you get more, you have to share that with those who don't know how to get it – not give it to them free. Don't give them the fish, teach them how to fish. One of my ways of giving after I've received so much is to show you how do it. If you give more, I guarantee you, you will get more.

We have covered significant ground in this book. We started at the core of the topic and visited the structure of the universe so that you understand that there is some logical rationale to all this. What you should see by now is that manifesting your dreams into reality is not a one-step process. It's not even something that you can do in steps. This is not like a pudding's recipe.

To manifest your dreams you need to live a lifestyle that aligns all these skills so that you can live the life of a conduit that can ask the universe for whatever you need, and it will flow through you into reality. Just like Thomas Edison made the light bulb the reality.

Asking how to manifest is an obvious question. But after reading this book, asking how to manifest is like asking a tuning fork how to play a note. If you follow a list, you may be able to manifest your dreams from time to time, or probably none at all. But when you change your mind, when you fine tune your vibration, and if you live life, then be careful, you will be able to manifest everything you want.

Because three things will happen: First, you can be inspired as to what are the things that you want. That means your desires will not be random and they will not be fancies of the moment. You will start to see your true value and the things that really mean something to you. When this happens, you will start to manifest effortlessly.

Second, you will start to have the energy to proceed with the action needed to make things happen because you will have the power to move anything out of your way, and climb over stumbling blocks that would ordinarily trip anyone else up.

Finally, you will be able to create the perfect vibration at all times. When you are at this level not only can you manifest constantly but things will also come to you automatically. They will be ready waiting for you, and they will stand in front of you.

Manifesting is about making your dreams come true and about making your life one that is meaningful and based on achievement.

For those of you who are wondering how to manifest that new Porsche, or the new Gulfstream 650, then I have some additional words for you.

Rewards vs. Achievement
When we look at monetary gain and comforts, there is nothing wrong with that. There is nothing wrong when you are motivated by the rewards of your achievement. Don't feel sorry about that and don't feel that you are a lesser person for loving the rewards of a rich life. But what this book is offering you is more than all that.

Anyone with decent credit is going to be able to get a loan to buy a Ferrari. In fact, Gulfstream even has financing. Or you could even get a time-share for jets. What's the big deal in that? What you should really focus on is the achievement because the reward will always follow. When you visualize the sports car in your driveway or the super-yacht in the marina, all you are doing is imagining the rewards that you want.

However, if you focus your powers of manifestation on making a million dollars a day, or ten million dollars a day, or even 100 million dollars per day, then what you end up doing is creating the vibration that results in power. With 100 million a day, the toys you want to buy will be done effortlessly.

The other thing is that if you focus on the reward and ask for a yacht, let's say. What happens after you get the yacht. What happens if you just scrape enough to put together a loan and the yacht is in the marina. What's next? The disappointment will be substantial.

Instead, desire the achievement of success. The rewards will follow.

I want to leave you with one last thing to consider. The universe only gives what you want – and that could be anything, even setbacks. If you desire bad things, that's exactly what you will manifest. When you are negative and you vibrate negativity, that's what you will manifest. You can even manifest catastrophes. This is why bad things happen to people with negative thoughts and that reinforces their resolve that bad things always happen. If you are someone who has a lot of bad things happening in your life, then you need to start thinking positively. Go back to Chapter 4 and look at the skills you need to practice to be able to think positively so that you can start changing your life. It only takes one event that you desire positively, believe that you deserve it and allow it to happen without hesitation for you to realize that you can have good in your life and it can be good after good. It's what you believe will happen.

Where Manifestation Comes From

Everything is energy. What we perceive as solid and separate is an illusion, a projection of sorts. Underneath that is a constantly flowing, interconnected field of energy that vibrates at different frequencies. Solid matter consists of energy fields on the quantum level which are vibrating at a particular frequency and interacting in a certain way. Matter may appear to be solid but it is constantly moving and vibrating at the sub-atomic level. Thoughts and emotions are also energy, vibrating at much higher frequencies and interacting with the surrounding field of energies. A thought of love or joy has a completely different frequency than a thought of despair or anger. Energy draws more of like energy onto itself, therefore positive thoughts of a certain frequency will draw more energy of the same frequency to them (positive events). Equally negative thoughts will also attract similar frequencies, which will result in compulsive negative thinking and corresponding events realizing.

You are an energy being with a physical experience. We in our physical bodies are translators of vibrations, we interpret different frequencies with our senses. You are energy and you are sending out a signal, a frequency that then attracts similar frequencies back to you. Thoughts and feelings are high-frequency energies that are not perceived by the senses, nevertheless, they exist. Your thoughts and emotions determine the frequency you are sending out and therefore attracting.

THE EMOTIONAL GUIDE

Your emotions are a wonderful and effective guide you can use on your journey of creation. Good feelings mean you are going in the right direction towards what you want. Bad feelings indicate you are off course. You can easily differentiate what you are focusing on by how you feel. If your vibration is matched to your desire you feel elevated emotions, you feel good, joyful, and enthusiastic. When you are focusing on lack, generally the vibration will be lower, resulting in lower feelings such as worry, anxiety, frustration, or hopelessness.

Your current physical reality is a reflection of your most frequent thoughts and feelings.

This may seem intimidating at first because most of us have not learned to consciously control and guide our feelings. The good news is, and perhaps one of the most important tools that you can use is that you can always choose how you want to feel. You may not be able to control how other people feel or act but you always have control over your emotions and thoughts. To achieve a more positive, or high vibration, feeling you have to simply think of anything positive, whether it is in your life now or you want it to be.

True freedom lies in being able to maintain inner peace, joy, and happiness, regardless of external circumstances. And you can achieve this, by choosing a thought that feels better, over, and over, regardless of what is going on outside of you.

Your desires and beliefs must be a vibrational match for you to receive what you want. You cannot want something, focus on the lack of it and then expect to receive it, as they are completely different frequencies. You must allow it to be received by you by tuning into its frequency,

Be conscious of what influences you allow to enter your experience. Excessive attention to news and media can lower your vibration, as these can evoke a lot of negative thoughts and feelings. Always reach for the best feelings. Watching programs that make you laugh and generally feel good are a much better choice than programs that evoke fear or anxiety.

Similarly, be conscious of the people you surround yourself with and also of how you interact with them. Do people around you always complain? Do they always talk about their problems? Do they gossip or judge others? These are all negative influences and should be kept to a minimum. Where not possible to completely avoid them, make a conscious effort to not engage in that kind of conversation as that evokes a lower vibrational state and makes you more susceptible to unpleasant feelings.

Do you tend to complain to others or talk about your problems? If this is the case for you, make a conscious effort to disengage from this type of behavior. Complaining about something or commiserating with others about your unfortunate circumstances only perpetuates the situation. As you talk about it, over and over, you literally relive and energize it, you focus your attention on it, and you feel the feelings and think the thoughts associated with it. Your vibration then matches that of your problem, which then continues to draw more of it into your experience.

Avoid comparing to others. This is a completely pointless mental activity that rarely brings anything positive into your experience. Everyone's journey is different and their own. Focus merely on your own experience and on your own ability to shape it into what you want. You have no control over anyone else's reality or experience, only your own.

Why do we want abundance? Because we believe it will make us feel better. Anything we want in life is because we believe it is going to make us feel better in some way. With the understanding that you can always choose how you feel, you can feel better right now. By offering this vibration, you will then match your desire. Therefore, the only manifestation you should be working towards is that of the feeling of your desire.

Manifest the emotion of your desire. Just focus on manifesting the feeling and sensory experience of what you want, and you can do that anytime, anywhere. Make that your primary objective and the physical will catch up.

You can easily identify where you are in the emotional range (diagram below) by the way you feel. This is the vibration you are currently offering. By reaching for a better feeling thought, you start moving up this range. You will notice this in a form of relief or feeling slightly better. Bear in mind you cannot jump from despair straight to joy. This is a huge vibrational difference, and it is unattainable. You move up the emotional scale gradually, by reaching for a higher vibration emotion, feeling relief, finding your footing, and then moving up again.

How to Manifest

Clearing

To manifest what we truly want we need to be clear. Every second the brain receives 11 million bits of information. We only process 15 bits of information every second. So what about the other 10.999985 bits of information? Those are just filtered out by the brain as not valuable for our survival. This analysis is different for everyone. So if we can do something about the filter that analyzes all the data coming in, then we can change the way it is processed, and we can make the filter let in other information and even let in more information. This is where we start to work with our subconscious mind, we need to get clear. The subconscious mind is cluttered with pre-programmed limiting beliefs built from our experiences in life, and in some cases also what has happened to our family and ancestors, stored in our DNA. For example let's say we want to manifest financial abundance. But nothing is happening. Maybe we have a program running in your subconscious that says that "money is the root of all evil" which we all have heard some time in our life's but this is a limiting belief. This limiting belief is blocking us from achieving what we want; there will not come any money. We need to clear this limiting belief.

Identify your limiting beliefs

How do I know what my limiting beliefs are? That is the question, we can't start clearing if we don't know what to clear. Or we can but if we know what they are it will become easier for us in the process of clearing them.

The best way of finding these limiting beliefs is to sit in meditation, visualize what it is that you want, if it's the expensive car that you want then to visualize that. Then you might notice what kind of thoughts and feelings are coming to you? Do you feel frustrated and sad, maybe you think "I will never have the money to buy that car" or "what would my father say if I had a better car than him" or "I do not deserve that kind of a car". There you go, write them down they are the limiting beliefs that are blocking you from having that car.

Another example: you want to be a good public speaker. You sit and visualize yourself speaking in front of people. You get a feeling of anxiety, go into that feeling and see what is coming up for you. Is it a past event that made you fear public speaking? Ask yourself, "that which happened before, do I want it to affect me now? Or can I let that go?" Maybe you think "I can't do it", "I am too nervous", "I stutter and blush". There you go that are your liming beliefs that you want to clear from your life.

How do we clear? There are different ways to clear limiting beliefs; in this book we will talk about four of them, gratitude in the now, EFT, "Do I believe that"-method and the *Ho'oponopono*.

Gratitude in the now

This might be the easiest way to clear there is. The magic is to feel gratitude for whatever you have in your life, a pen, your children, your house, a flower. Just feel the gratitude at the moment, right where you are. To start your day in this way, your day will just get better and better. So whenever something is troubling you or you feel a negative emotion that you don't want to have or you feel is not in line with the situation, use this tool, feel gratitude for something in your life to stop the negative emotion from having any power over you. For example: you hear something troubling on the news, you get a lump in your stomach, just notice the feeling, don't dismiss it, just feel it, and then bring in something in your mind that you are grateful of. You will feel the shift so strong.

EFT

Emotional Freedom Technique. This is often used by life coaches because it is one of the most powerful tools to release stress. The technique combines affirmations and a specific tapping technique. The tapping is made with the top of the fingers at certain point on the body.

Before you begin the tapping assess how intense the negative emotion is, from one to ten. When you are done you can assess again and see how effective it was on that specific emotion.

First, you tap on the outside of your hand while you say: "Even though I have this *blank* (fill in the problem you have a limiting belief, negative feeling or something else) I deeply and completely accept myself". So for example:

The important step is to feel how this problem makes you feel when you do this. You say these three times while you tap the first tapping point (outside the hand). You don't have to say the whole phrase when tapping the other points, then you can say: "This limiting belief" or "That money is bad", "This sore shoulder", "This fear of spiders".

The tapping point is:

1. Outside of the hand (karate chop)

2. Eyebrows (both side)

3. Side of the eyes (both side)

4. Under the eyes (both side)

5. Under the nose

6. Chin

7. Beginning of the collar bones (both sides)

8. Under the arms (both sides)

9. Finnish at the top of the head.

This ancient technique became well known in modern times through Dr. Hew Len and Dr. Joe Vitale, and if you like to look into this deeper there are two books I recommend: *"Zero limit"* and *"At zero limit"* both written by Dr. Joe Vitale. You will find them in the recommended books under extra materials.

Anyway, this technique assumes all problems around you are created by you and you are therefore responsible for them. So when you have a problem in your life, or if you hear of a problem someone else has that makes you feel troubled, just feel the feeling you get when thinking of this problem and say four phrases. Say them over and over out loud or in your head, until you feel at ease again.

The four phrases are:

I am sorry

Please forgive me

I love you

Thank you

When the negative feeling is gone, it is cleared. You can clear on everything. We don't know what is in our subconscious mind, so we need to clear whenever we feel judgmental, troubled, or have any negative emotion, use the four phrases and clear. Try it out now: I am sorry, please forgive me, I love you. Thank you, do you feel it. It works and it works fast.

"Do I believe that?"- Method

A limiting belief has power over you as long as you don't see it and as long as you think that you believe it. Most of our liming beliefs are hiding in our subconscious and control us without us even thinking of them. Then it's very helpful to bring them into our consciousness and break them down. If you want to use this method to get rid of your limiting beliefs, then you can follow the following steps:

1. Pick your limiting belief, for example: "I am not worthy of that car"! "Money is the root of all evil", "I cannot speak in front of others"

2. Ask yourself: Do I really believe that thought? If your answer is No! Then you are done, the limiting belief is broken, and gone. If you don't believe your limiting belief, then it has no power over you anymore.

3. If your answer is Yes! Then ask yourself "why do I believe this?" Maybe your answer is:" because I am a bad person". Then ask again "do I believe that I am a bad person"? Again if the answer is No! You are done.

4. If you answer Yes then keep on going "why do I believe I am a bad person". In the end you have broken the limiting beliefs down.

That was the first key: clearing limiting beliefs. Now let's continue with the second key.

Feel Your Intention
When we know what we want in life, what we want to manifest. For example, the love of your life, a new house, a new job, a car, a better relationship with your children. Then we set the intention. How do we do that?

You take a piece of paper and you write down what you want. In so much detail you can. You only have to write it down. If it's a car that you want to manifest; then you write the brand of the car, the model, the color, how it feels to sit in the car, how it smells, every single detail you can think of, write it down.

Then find a quiet place and sit in meditation. Now visualize that what you want, let yourself see you sitting in your car, drive it, and see all of it in your mind.

Here is the secret, don't just play the vision in your mind. You have to feel it. Feel how it feels when you have the thing you want, how does it feel in your body when you are driving the car of your dreams, how does it feel when people look back at you when you drive past them? How does it feel to touch the car? The trick is to feel it as if it has already happened, pretend that you already have it. You can even play it in your mind as it was something that happened a week ago, or a year ago. The thing is that we want to trick the brain into thinking that this has already become true, you already have the car.

Sit with this vision for five to ten minutes. Then comes the last part of the intention setting, we have to LET IT GO!! You don't have to control the outcome, don't have to force it and you don't have to know how it is going to manifest in your life, just let it go to the universe, and let the universe answer to your desire. That takes us into the third step inspired action.

Take Inspired Action

To make your intention come true you can't just sit and wait for it to land in your lap, you have to take inspired action. When the universe responds to your intention, you have to answer the call. This is to take inspired action, it is like you set a request to the universe and then the universe guides you on to what to do and when to do it, to get your desired intention filled. When we take inspired action, we are answering the call from the universe. It can come as an inspiration to do something, to contact someone, to read a book, listen to a podcast. Or you just feel an urge to do something, this is when you act! The important part is to keep an open mind and to be vigilant for the signs. Whatever you feel your attention is drawn to, make sure you explore it; it could be the call from the universe.

If you take inspired action you will end up having your desired intention sooner than you think.

These are the secret keys to manifestation. Start using them! And the life of your dreams is just around the corner.

How to Raise Your Vibration

Someone once told me that she did not dance because she was happy; she danced because she wanted to get happy. The simplicity of that concept is amazing. It's a consequence of a similar statement regarding an external expression of inner grace.

Both these aphorisms brilliantly show you the way to change your vibrational state. If you are feeling blue, for instance, you're going to attract some unwanted situations into your path. To avoid that you have the option to consciously change your vibration by doing an external act to change your vibration. In my friend's case, she chose to dance.

The state of attraction is not just about attracting good fortune and material gain, although it can be. However, using something so powerful exclusively to gain things that don't mean much from a cosmic perspective is a waste. There is so much more than you can do with the law of attraction than just that.

Stepping up to the state of attraction requires the invocation of a state of peace which you can attain from balancing meditation with physical activity. You can advance that pursuit by also doing many outward actions that bring about internal changes in the state of your vibration. This along with meditation, mindfulness, and exercise, elevates your game. But there is more. After all, the universe is abundant in its resources and there is no limit to what you can ask for or what you can achieve.

Rituals

To fine tune your vibrational state you can add one more outward expression to an inward grace and that is the execution of rituals.

Rituals are directly connected to the law of attraction in the sense that when you believe a certain ritual brings about an event, you instantly resonate at the frequency that attracts exactly what you want.

Rituals are not ancient mumbo jumbo that have lost their meaning in contemporary society. Science has developed methods to research and test the effect of rituals and has come up with statistically relevant evidence that rituals have an effect.

Don't think of rituals as something you find in ancient religions. Every religion, even in the modern sense, has some form of ritual. Look past these religious foundations. Look instead at the coincidence of success in the wake of performed rituals.

Take Alexander the Great, for instance. He hardly lost a battle during his conquest of Asia against the significantly larger Persian forces, and the smaller provincial armies as he trekked to the heart of his adversary's empire. Each account of his battle is more amazing than the earlier one but illustrates his ability to overcome great odds.

It is well documented in history, how he preceded each campaign with rituals.

Genghis Khan did the same. Even Joe DiMaggio had his own ritual before every game, as did Wade Boggs and every other major league player.

Rituals are like the tuning fork of nature. They help, in concert with being in a state of attraction, to amplify the frequency required for effective attraction.

However, do not get carried away with rituals. Some people become obsessed with it and that state of being in obsession alters the vibration and diminishes any potential effects of attraction.

A good way to keep yourself in a state of attraction, where you are constantly buzzing silently at the right vibrational frequency is to have a daily morning routine. This routine should include meditation, exercise and a simple ritual that gets you into the right vibe.

Fine tuning your state of attraction takes practice and discipline. Both of which need to be slowly cultivated and developed. This vibration is best taught to children so that it becomes second nature to them. It is also easier for kids because they have less mental mess to deal with it. For adults, fine tuning the state of attraction to the point that it is always in the state of attraction will have a tremendous impact on their efforts in life.

Scripting
The first thing that would come to mind when the word 'scripting' is mentioned is acting, drama, or

plays because, of course, actors often have to memorize a set of scripts written by scriptwriters when acting in a movie.

Scripting is no different only that this time, you are both the scriptwriter and the actor.

Scriptwriting is a simple technique that involves writing out the story of your life the exact way you want it to turn out or writing about a specific thing that you want to happen to you in the particular way that you want it.

It is writing out what you are trying to manifest as if it has already come to you.

Scripting is like playing make-believe; you are writing a story of your life but you are writing as though you are already enjoying the benefits of what you want and feeling the vibration as if you already have it.

You paint this elaborate picture and write this elaborate tale of how you want something to manifest, that thing being maybe your life, perhaps your day, your relationship, your career, your coffee, the sky is the limit.

It is very similar to writing your intentions, only that this time you're tapping into your imaginations and making a creative story.

Creative Thinking to Improve Your Success

Creative visualization is a technique in which you use your imagination to make your goals and your dreams to life. If you use creative visualization techniques in the right way, you can greatly improve your life and your chances of prosperity and success. See it as a kind of super-power, one that you can use to change the environment you are in, maybe the circumstances and make things happen. It can be used to attract anything you want into your life – people, love, work, and money, whatever you want.

Visualize in your mind an event you want to happen or an object that you want, and you will attract it to you. Creative visualization is somewhat similar to daydreaming and, while it might look as if it is magic to some people, it is not. The only thing that is involved is the power of thought and natural laws of mentality.

Some people use this in their everyday lives, not aware that they have this power and are using it. Anyone successful has creative visualization to thank for it, whether they are aware of it or not, purely because they visualize what they want, they see the goals they have set as already being accomplished.

How does it work and why?

That is an interesting question. When you repeat thoughts, think about something more than once, your subconscious picks up on it and starts to accept these thoughts as reality. When it does that, your mindset changes along with your actions and your habits. Making those changes will bring you to other people, other situations, and places, molding your life and attracting the very thing that you are thinking of.

Thoughts can travel from one person's mind to another. If your thoughts are strong enough, they may, unconsciously, be picked up by another person who may well be in a position to help you realize your goals. You see, thought is a kind of energy and that energy is immersed in emotion. Because of the energy that your thoughts contain, they have the power to change the balance of the energy that surrounds you.

How often do you think the same thing over and over again? Quite a lot I would imagine; most people do. Whether you are conscious of it or not, your thoughts are centered on the situation you are in and the environment that surrounds it. Therefore you can create the same environment over and over again.

When you watch a film you think you have seen it all. Now watch it again but this time, change the way you think about that film. You will find that you will view the film quite differently and will pick up on many things you didn't get the first time around. You can create a different kind of reality.

Again, this isn't magic or supernatural power; it's just pure natural power that we all possess. It's just that not everyone is aware that they possess these powers and perhaps don't use them as they should be used.

Overcoming Limited Thinking

Creative visualization can create great things in your life but we must remember we are not all the same. We cannot all change the same things about our lives, at least not straight away, so you must be aware that, although you have this great power, it is somewhat limited, but these limits are inside of us, they do not have that power. Most of us limit ourselves at times and cannot see beyond a certain area, and that limit is placed by our thoughts and beliefs. In short, we are limiting ourselves to what we know, not to what we could become.

The key to creative visualization is being open-minded, thinking far bigger than we ever dared possible. The bigger you think, the more opportunities come your way, provided you can rise above the limitations in your mind. Don't expect instant results – it will not happen overnight. You may see some small changes to start with but, the more open your mind, the more you think big, the more will come your way.

Two keywords for you to remember – patience and faith.

Living an Abundant Life

You can attract wealth by using the Law of Attraction, and just as in looking for love, your thoughts and vibrations need to be positive if you are to achieve the wealth you want in your life. When it comes to money, you can prepare the way to remove your negative thoughts by consciously becoming more proactive in managing your finances. Think of money as your friend and visualize yourself with all the money you need. Imagine having that money in your hands and imagine what it feels like to have all the money you want. In the meantime, you can take steps to move yourself towards that state.

The most important thing you can do is to arrange to pay your bills on time, so that you are not worrying about debts. This creates negative vibrations, because when there are obligations people cannot meet, whether financial, professional, or personal, they become fearful. Not only that but you also cannot truly focus your thoughts on making money if you owe money. The negative cancels out the positive here.

Alongside this, list all your expenditures so you can identify areas where savings can be made, and create a realistic, workable budget. If all this seems like a lot of hard work, think of it as removing the negative money-related aspects from your life. You are preparing the way for abundance by being proactive in reducing debt and managing your money wisely.

Remember to be thankful for what you have, rather than wishing for what you do not have, because those are negative thoughts, and they will return negative manifestations. Now you are ready to use the Law of Attraction to attract wealth and abundance to you and your loved ones.

The first thing to realize is that wealth does not relate exclusively to money, and you can be wealthy in your life even if you do not have a lot of money. This is where being thankful comes in. If you are desperate for money – for example, if you focus your thoughts on winning the lottery to solve all your financial problems – then you will repel money and attract more desperation. It is all those negative thoughts surfacing again.

One thing you need to remember is that you get exactly what you ask for. So, if you ask to be a millionaire but don't ask for the things you might want to buy with that million dollars, you will not get them. What you ask for, you will get, so be specific in your requests. Here are some ideas to help you maximize the potential for attracting money.

Shift Your Focus Regarding Bills and Debts
It's a natural response to complain when the bills come in, and of course, those are negative responses. 'How did we run up such a big bill on the credit card?' is not useful. Telling yourself that you are grateful that you used the credit card to pay for a new washing machine so that you can keep the family's clothes

clean with ease demonstrates gratitude. Gratitude for a credit card bill might sound a bit surreal but that's the way you're going to have to program your subconscious if you want to use the Law of Attraction to attract money.

Think of the things you have been able to do as a result of using electricity, for example. You've cooked food for the family, been entertained by the television, and had light during the hours of darkness so you could read, work on the computer, or catch up with friends through Facebook. Put it like that, it's almost a privilege to have that electric bill to pay, isn't it?

Realize How Much Abundance There Is Already in Your Life

Okay, maybe you are not a millionaire – yet – but you are rich in other ways. You have a roof over your head, food to eat, and an abundance of friends who care for you. Just get into that mindset that everything is something to be grateful for and that you have an abundance of good, useful, and beautiful things in your life, and you will soon attract more abundance, because, as you know, like attracts like. And the type of abundance you will attract will be what you want, what you have asked for, whether that is money, possessions or even both. If you consider yourself to be rich in non-financial ways, then eventually you will be rich in money and possessions. It's all about reprogramming your subconscious to open your mind to the reception of wealth and abundance. If you think negative thoughts, then it will attract negativity. So don't think you don't have enough or that you could have been a billionaire if it weren't for your fate. You have a lot now and you have to feel blessed for having all of it. If you think what you have now is a lot, then wealth and money will keep coming into your life.

Love Money
The old saying that 'Money is the root of all evil' is not true. Money itself is not evil – it is a thing, not a person, it does not have characteristics or a conscience. It is the way some people use money that is evil, not money itself. So there is no reason not to love money – it can do a lot of good, and if you have money, you can do a lot of good with it too, for yourself, your family, and those you love.

Spend money – enjoy spending it and enjoy using the goods and services you can buy with it. Build an affinity with money – love the smell of it and the touch of it. Love what it can do to improve your life. You have to fall in love with it.

To attract money into your life, you have to think of it in a positive way, so that you can radiate positive vibrations and attract that which you are thinking about. Let go of your negative thoughts about money, bills and debt and celebrate its existence. On the other hand, don't be greedy or miserly. These are negative emotions that will repel the wealth you are seeking.

Above all, don't envy the success of others – it is a fruitless emotion and it will not result in success coming your way. If you criticize or envy their success, it will not help you to draw success to yourself. In fact, envy is a real barrier to receiving abundance.

15 Best Secrets of Manifestation

To manifest your desires and dreams, you need to first relax your mind, and align your mind, body, will, heart and soul. If your entire self is not aligned consciously with your desire, you may delay or may not see the desired results. In fact, you may experience resistance and counterproductive results in the end. So focus! That's your primary key to success. You can manifest anything you choose, whether it is earning a lot of money, buying your dream house, restoring health, or having a successful and loving relationship. The key is, believing in your dream and believing that it is possible for you. As William James said, "The greatest discovery of my generation is that a human being can alter his life by altering his attitudes."

It is important that to get to your end goal, you must stay calm and relaxed, and work towards it using physical, emotional, and spiritual strength and effort together. Your dream will become your reality!

• Never use the words "I can't" – Murphy's Law states that "Anything that can go wrong, will go wrong". Think for yourself instead of allowing Murphy's Law to govern you and your life. Always remember, where there is a will, there is a way. So if you want something, there is always a way to get it. Relax your mind, immerse yourself into your dream and live it as your reality. Your positive energy will magnetize the Law of Attraction and manifest your true desires and intentions. Negativity will obstruct your desires and mess up your manifestation energy. So keep your mind calm and focus on your desires. Nothing is impossible if you have the will to make it happen. You are not alone because even the Universe wants to help you.

• Stand tall with your head held high – You need to be proud of who you are. You need to keep your inner-self calm and maintain a good foundation. When you dress confidently and keep a positive outlook, you will exude even more positive energy. This will further attract positivity and abundance into your life every moment. You are the Universe! You can tap into the abundant source of energy in you and connect with the Universe quickly. When you believe in your dreams, you strengthen your manifesting energy. When you allow stress to control you, you will only mess up your manifesting energy. Minimize negative emotions from your life. Start stepping into your new life. Dress as if and live as if you are already the person you want to be. Stand tall and proud. Feel the success. Make a list of your positive and negative traits. Now actively work on eliminating your negative traits and strengthening your positive traits. See how you will be filled with high levels of positive energy within you. This will not only give a boost to your personality but also help you manifest your dreams quickly.

• Visualize your reality – Remember the basic principle behind manifesting – thoughts become reality. As Buddha said, "All that we are is a result of what we have thought." To create what you want; you need to visualize it. Visualization is also an integral part of the creation process. The more clearly you see your desires, the quicker you will attract the Universe to manifest them. Dedicate some time for yourself so that you can spend time to relax and focus on creative visualization. Visualization is also a powerful tool to eliminate all negative vibrations. One of the easiest ways to visualize it is by relaxing yourself completely. You can relax during a hot bath, or by listening to soft music. Do anything that can calm your body, mind, and inner spirit. Immerse yourself in the moment. Experience the moment. The more you focus on your creation, the more you will generate positive vibrations and this, in turn, will energize the Law of Attraction to manifest your dreams into reality. As Albert Einstein said, "Imagination is everything. It is the preview of life's coming attractions."

• Embrace good vibrations and generate positive energy every moment – The moment you start thinking positively, you automatically emit positive vibrational energy. You can emit positive vibrations in your routine activities like talking, visualizing, feeling, smiling, thinking, etc. Do you know that you are constantly emitting vibrations? Whether they are positive or negative is up to your thoughts and actions. If you are constantly worried, tensed, scared, nervous, etc., then you are generating negative vibrations. This can mess with your manifesting energy. On the other hand, if you relax and experience inner peace, you will start emitting positive vibrations. This can magnetize the Law of Attraction into manifesting your desires. If you want to change your current circumstances, you need to shift your energy to what you focus on, and what your dominant thoughts are. That way you start manifesting from a higher energy vibration and your desires turn into reality quickly and easily. The moment you are relaxed, happy, and confident, you will emit powerful positive vibration and allow everything you want to flow to you effortlessly.

- Create a list of "Must Haves" – You should understand that each step you take brings you closer to your desired outcome. So first and foremost understand your true needs, desires, and intentions. Unless you understand and believe what you truly want, you will not be able to focus on them. This can result in negative vibrations and ultimately delay the response from the Universe. To be able to use the Law of Attraction efficiently, first relax and ask what it is you truly want that will bring happiness and prosperity into your life. When you are in a state of tranquility, you will be able to connect to your inner self that can guide you into making the right decisions for yourself and your future. Be specific. Refine. Once you know what you truly want, focus on your goals, and take concrete steps towards achieving them with the help of the abundant source of energy of the Universe.

- Love yourself – You need to always love yourself first as that will give you enough confidence to make all your dreams come true and remain focused till they come true. You need to invigorate yourself with the abundant source of power deep inside you. To tap into that eternal energy, you need to practice meditation or deep breathing. This will calm your body and mind. You will become more self-aware. Now you can connect to your true inner self. When you connect to your true self, you will find abundant happiness and confidence that is stored within you. Now focus on your goals and consciously use the Law of Attraction to manifest them. It is so easy to make your dreams come true. It is time you try it and experience the life-changing opportunity.

- Give your best – When you want good things to happen to you, you need to first give your best. To experience the best in life, you need to first present your best. There is no shortcut to success. Whether it is finding true love or getting your dream job, you have to work hard to manifest your goals. You have to first calm your body and mind, focus on your goals, and lay concrete steps towards achieving them. You cannot rest until you attain your desires. Don't be disheartened! Plant the seeds of your desire today. When you stay focused and work diligently, the laws of the Universe will come together to help you reach your destination without fail. You have to take the first step every time!

- Clear your mind – Before you want to manifest, it is essential you first clear your mind. You can practice deep breathing or meditation to help clear the mess in your mind. Drive out all negativities. Let it float away without any effort. Once the slate is clean, you can start building your reality. Fill the cup with positive thoughts. Remember, if your thoughts are vague, the results will be too. Be clear about your wishes and dreams. Don't be afraid to ask for exactly what you want. Try to feel how you will feel when you achieve your goals. Will they make you happy? If not, clear your head again. What do you want? When you connect to your inner self, the Universe will guide you to make the right decisions. Unless you feel emotionally connected to your desires, keep thinking till you know exactly what you want. The best way to do self-introspection is to meditate.

Connecting to your true self is easier than you think, and you will find answers to most of your questions deep within because the Universe can respond to your questions in a state of tranquility. You can understand the vibrations easily when you are calm. You will not be able to do it any other way. Relaxation is truly a powerful tool for self-analysis and understanding. Meditate to find answers to your problems. When you are stressed, take a break from work, and calm yourself. After a period of stillness, when you again look at the same problem, you will miraculously find an answer you have been searching for so long.

- Manifest your desires daily – As Winston Churchill said, "You create your own universe as you go along." Give some time to yourself every day. It can be anytime you want or anywhere you like. Just relax. Take four deep breaths and focus on your goals. When you are constantly thinking, you are constantly manifesting. You can also manifest through your routine activities, such as thinking, feeling, loving, touching, etc. You can exude positive vibrations through many activities. You need to live in your dream as if it is already your reality. This way your manifesting energy will magnetize the Law of Attraction into making your dreams turn into reality. All you need to do is think about it, feel it, and believe it!

- Relax – Relaxation is a powerful tool to manifest. To fully focus and consume yourself with your dreams, you need to first calm your mind and body. Connect with the abundant source of energy within you and use it to create strong positive vibrations. When you manifest from higher energy levels, you will reach your goals quickly and easily. When you relax, express gratitude for the fulfillment of all your dreams. You need to believe that your desires are already on their way. If you already believe that you have it, you already have it. The Universe will work together to make your dreams come true. Just sit back, relax, and watch the magic of your dreams turning into reality in no time.

- Be positive – As Charles Haanel said, "The Predominant thought or the mental attitude is the magnet, and the law is that like attracts like. Consequently, the mental attitude will invariable attract such conditions as correspond to its nature." What you think and feel is what manifests into your reality, so you must not have any negative thoughts or emotions. Fill yourself with positivity. If you tell yourself that you will succeed no matter what, most likely you will. When you emit positive vibrations, you will attract more positive energy. So just relax your mind and think positively. Good things will happen to you. Make a list of your negative traits and try to get rid of them as fast as possible. Then you will be ready to attract abundance into your new life with the magic and power of the Law of Attraction. As Buddha said, "Your worst enemy cannot harm you as much as your own unguarded thoughts". Don't let negativity harm you anymore. You have unlimited chances to change your life for the better. Make an effort to shift your focus to more positive thoughts and watch miracles unfold before you.

- Live each moment to the fullest – Every moment of your life, you are constantly manifesting with the vibrations produced by your body. You know best whether they are positive or negative. If you relax and calm yourself, you can eliminate negative vibrations and only have positive vibrations. When you manifest from higher vibrational energy, you will likely reach your goals faster. It is essential you do everything to keep yourself happy. Create a bucket list and start doing them right now! Fulfill your desires right away and let them increase your positive vibrational energy. Pamper yourself! Go out with your friends or family. Get a massage. Go to the gym. Take some time to laugh! Treat yourself to good food and chocolates. You have the right to be happy every moment. When you are happy, you can focus on your goals better and manifest them using the Law of Attraction. Everyone needs a "me-time" to do the things you love.

• Connect with your spiritual self – When you are calm and relaxed, you can connect to the spiritual inner self. This will increase your self-awareness. You will be able to connect to the Universe and directly communicate with the help of your vibrations. This will help you attain inner peace and happiness. You will emerge as a positive, happy, and confident person. You need to experience it first in your mind to understand how it feels. You can then start seeing manifestations of physical realities – wealth, career, love, and health – reflecting your relaxed state of mind.

• Experience the power of gratitude – The Universe thrives on love, gratitude, and kindness. This will help attract positive energy faster. Start your day by thanking the Universe for everything you have today. Relax your mind and express your gratitude for fulfilling your dreams. Believe that your desires are already on the way. When you are filled with gratitude, you automatically connect with the Universe and the Law of Attraction. You attract more abundance into your life. Be thankful for all the good things happening in your life today, no matter how small they are. Even a silent thought or expression can emit positive vibrations that can be heard by the Universe. This is the essence of the Law of Attraction and one of the reasons it can work on a person. If you want to manifest your dreams, don't beg but only be grateful.

- Overcome your fear, doubt, stress, anxiety – It is normal to fear the unknown but don't allow it to cause stress and worry. Don't allow negative emotions to consume your present. Life is beautiful so enjoy every moment of your life. We can control our thoughts, emotions, and actions. You have to trust the Universe. Relax your mind and body and surrender yourself to the abundant source of energy in the Universe. When you harbor fearful thoughts, you will emit negative vibrations. This can impact your life negatively. You will mess up your manifesting energy. Don't harbor any negative emotions like doubt, fear, desperation, worry, depression, etc. You will emit negative vibrations, and this will result in counterproductive efforts or failure. Don't allow any negative thoughts control you, especially when you want something badly. Instead, believe in your dreams completely and live them as a reality. When you start thinking positively, you will attract the Law of Attraction and manifest reality just the way you want.

So what are you waiting for? Start manifesting today. Just relax, be clear and remain focused. Exude a confident energy of everything going your way and watch how your energy in turn attracts your goals one step closer to you every day. As Nikola Tesla said, "If you want to find the secrets of the universe, think in terms of energy, frequency and vibration."

True Freedom and Lasting Happiness

Relying on faith is either a stopping point in our evolution of consciousness, or through the invitation of our spiritual will, the soul steps forward as the new source of greater truth and sense of identity. Consequently, unlocking an entire level of existence never before considered.

While reason and understanding are the highest themes of physical reality, it is ultimately losing access to true freedom and lasting happiness that come from acknowledging a spiritual reality. Without a spiritual context of reality, the logical mind is limited, and not able to enjoy the bliss of trusting in a Divinity beyond itself. Thus it will always be haunted by fear of death and the unknown of going beyond the physical. All themes of consciousness previous to a Spiritual Reality could be described as seeing the body and mind as the source of consciousness (existence), whereas the themes past this point view consciousness (existence) as the source of the body and mind. Consciousness is beyond time and space and who we are, our true Self, while the human body and mind are passing expressions of it. The perceived universe is more so being projected from within us, rather than something we are separate from. The universe we perceive only exists within our consciousness and is relative to our theme of consciousness. We all are perceiving different layers or dimensions of reality.

From a greater, non-dualistic perspective, even the notions of "inner" and "outer", "this" and "that", or subject and object begin to merge and eventually are realized to all be the same. Such a realization is beyond reason and leads into the realms of the mystical and nonlinear.

As much as it resists the unexplainable power of belief, the mind is forced to surrender to the fact that belief and faith dictate our experience of reality. This power of belief is most obviously demonstrated through the placebo effect and its defiance of reason and understanding. The placebo effect is the notion of giving a patient something that has no actual medicinal value but the patient's faith and belief in its healing abilities physiologically create an actual healing effect. This is just one of countless examples of the miraculous which demonstrate that belief and faith trump reason and understanding by methods that it cannot explain.

At this incredible tipping point, consciousness—our unexplainable spiritual essence/soul—has officially been accepted as our identity and we loosen our attachment to the external world. The realization of our spiritual nature brings with it exquisite, unlimited access to love and peace that can transcend physical circumstances. While the human ego lives in constant fear and lack because of the fear of physical survival, the spirit lives in the realms of true freedom and lasting happiness because it is eternal, and survival is therefore already guaranteed.

Love & Inner Peace

Archetype: The Spiritual Seeker

Triggers: Conditional love

Root Program: I am the source of love and peace

Transcendence: Unconditional love

With the dawning of spiritual awakening, emerges the opportunity to access love and peace from the eternal spring within. Inner peace is experienced as the natural consequence of finally remembering the truth of our spiritual nature and finding gratitude beyond circumstances. Therefore if all subjective states are an inner choice of perspective, it would only make sense to choose to see our existence from a perspective that evokes happiness, gratitude, love, and fulfillment. And as we align with our highest version of lovingness it only benefits ourselves and others, therefore only perpetuating itself.

It is worth noting that what the world considers love is typically referring to infatuation, possessiveness, pride, or wishful attachment. True love is embodied and is a lifestyle approach. As we move into the frequency of love we become aware of the love and goodness that is all around us, we just didn't recognize it before. True love does not need a reason to love, it just loves because that is its nature.

In the frequency of love and inner peace we start to finally experience the highest potential of reality.

Everything can be perceived with love if we stretch ourselves beyond the linear mind's parameters. It is as if we were seeing the world for the first time, and in many ways we are. Colors appear more vibrant, we notice the damp leaves glowing in the sunlight as if revealing a glorious painting, bumblebees are no longer a nuisance but instead become messengers giving us their blessing. We notice recurring numbers sensing they carry meaning and license plates stand out giving us signs. Nothing is just mundane or ordinary anymore. The trees are now dancing for us, the flowers are blooming just to show us how beautiful they are. Everything seems to be in synchronistic harmony. Everywhere we look we see the magic of Divine Love and life is forever transformed into heaven on earth through the newly discovered lens of love. The irony is the world did not change; we did.

In the frequency of love, everything has meaning and purpose and is communicating guidance to us, including our current situation and circumstances. Every challenge, or life situation, becomes an opportunity for further spiritual awakening. Every perceived setback becomes a redirection by Divinity. There are no more accidents or coincidences. Everything becomes a part of the cosmic dialogue between us and the Creator. Love perceives a loving universe, and everything can be viewed from a positive angle. Even when the meaning or purpose of an event in our lives is yet to be understood, we easily give it over to hope and faith trusting that it will be revealed to us in due time.

Because consciousness is vibrating in the frequency of love, it sees love everywhere it looks. As mentioned before, even debt is not an enemy of love. Love thinks, "Thank you God for providing me a way to get ahead and for providing me the way to pay this back tenfold. My future must be really bright!" If it has perceived competition in work or confrontations in the world it thinks, "Thank you God for providing me the opportunity to surrender, forgive, and exercise my faith in goodness. All is as it should be. What will be, will be. If this opportunity wasn't meant for me, it only means something even better is coming."

From the perspective of love we realize that all beings are an embodiment of a theme of consciousness and therefore what others bring forth only demonstrates their current nature, not our own. Hence the teaching, by their fruits you shall know them. If all experience of life is based on an internal narrative, then that means that nothing is intrinsically objective or personally offensive in life. For example, even when targeted by anger, hate, or slander it is only a projection of another's internal reality. All suffering is self-created through limiting root programs. Hence all great spiritual teachers taught, pray for your enemies, because they bring themselves down by their own hand.

These realizations allow us to not take the suffering of the world personally, which would just make us fall back into limitation ourselves. Instead we have immense empathy, compassion, and patience for the shadow side of humanity. Love respects everyone's own unique journey through the darkness. To any perceived adversary love says, "Thank you for the opportunity to forgive and learn about myself. How can I help you out of your suffering? I see you. I understand you are hurting. I love you." Thus compassion towards all life becomes the natural response in this theme.

There is nothing love cannot find gratitude for. Life is a gift and most of it is out of our control. What we do seem to have control over though is our perspective of this gift of life. Therefore the most loving perspective to take is that all is exactly as it should be, is unfolding for the greatest good, and that there is no need to judge or change anything. All is perfect as it is without accident, mistake, or coincidence. There is perfect justice, balance, and harmony that we may not be able to see with our limited human perspective but we trust in the perfect consequences of Creation that may stem vastly beyond this lifetime. Through believing and having faith in such perspectives we notice dramatic increases in our overall levels of internal freedom, joy, and inner peace.

All of love's existence is beautiful, hopeful, and based on radical faith. Doubt, worry, fear, anger, expectations, regret, pride, and greed are all things of the past because they are no longer attractive as viable options. In love there is enough to go around. Love is not concerned with taking but with giving. In this state of consciousness there is no longer fear of shortages. And love wants to show graciousness and abundance to anyone suffering from this fear in hopes that they too will come to realize the generosity of the universe.

Love moves beyond the physical realm and perceives energy and essence through intuition. Before the transition into spiritual reality, we might focus on linear details such as words, actions, facial expressions, voice inflection and tone, body language, or even someone's appearance when interacting. But when love interacts with another being it is primarily experiencing their subjective energy or essence (theme of consciousness). It is the difference between seeing versus intuitive knowing.

To go from conditional to unconditional love we have to expand our awareness of love to even include all that we would typically deem "unlovable", "tragic", or "ugly" in the world. This final step into true freedom and lasting happiness moves beyond preferences altogether. Instead, we explore the innocence of consciousness—which includes everything in existence. Unconditional love sees that our preferences are not better than our not-preferences. All things are equally divine for simply existing.

Unconditional love is only realized by letting go of the mind's attachments to dualities. "This" versus "that."

Lovable versus unlovable. Thus it is not that we develop the ability to love the "unlovable"—we simply let go of the notion of unlovability in the first place. All labels are fallacious and coming from an arbitrary perspective. All is perfect as it is, beyond the labeling.

In a unified, interconnected universe, all is exactly how it is supposed to be without possible mistake, or error. To resist otherwise is to fight the entire universe being what it is. The limited ego resists love and focuses on all the proof for how unloved it is, whereas spirit focuses on all the proof for how loved it is. Both perspectives will technically feel "right" and justified but the experience and consequences of each will be different. All humanity could be in absolute joy of existence in an instant if we accepted this effortless choice. For it is absolute joy to live knowing all is perfectly aligned by Divine Love and not a single hair is out of place. All we must do is to accept it. Everyone has access to Divine Unconditional Love.

True freedom and lasting happiness are the feelings of finally returning home to full security, joy, love, peace, gratitude and ineffable awe. Whatever it took to realize the truth of Divine Unconditional Love is seen as irrelevant and worth the cost. For, if that's what it took to bring us to this moment now, then it was perfect and exactly what our consciousness needed to awaken. All of us are on our perfect journey of remembering just how loved we are by Divinity. To exist is to already be loved beyond comprehension.

Yet, we seem to enjoy the hero's journey through the relativity of not knowing. Thus we explore all that is of our not-Self only to return to what was always there, we just didn't see it yet.

Unconditional love compassionately understands that some souls need to wander into dark confusions of shame and guilt to prove that they are still infinitely loved, while others accept it more quickly and avoid the temptations of the ego's defiance of Divinity. Either way, it is all perfect and only revealing what is eternally available to all. However, Divine Love cannot force itself and can only come when invited. Hence this writing is an invitation to true freedom and lasting happiness.

30-Day Plan to Raise Your Vibration

Now that you've completed all of the previous chapters, you may still be thinking that you are not sure how to regularly put this into practice. A great place to begin is by setting an intention to do this for 30 days and be sure that you create a new habit.

You read the daily routine earlier but before we even focus on how to daily select your affirmations, it's a good idea to pull out again your list of goals and see if you can find patterns in your goals and create topics or categories. Most people find that their goals center on their health, from emotional to mental to physical wellbeing, or they center on finances/jobs/wealth and relationships. Sometimes you may have another goal like education. Find your categories so you can begin categorizing your needs as you make your lists.

Before you proceed to the 30-day plan, there are a few things you need to know:

- This plan is designed to work for both beginners and advanced users of affirmations

- The plan is meant to be adapted to your specific goals

- Each week of the 30-day plan contains a focus

- There is an optional advanced section that you can add to the plan for those looking to step up the challenge

- Each day revolves around a morning, afternoon, and evening repetition

- Morning is focused on who you are, afternoon on what you have, and evening on what you will do.

The 30-Day Plan

Each week is split into an area of focus. You are encouraged to change focuses with ones that match your desires and goals. There will be additional focuses beyond the ones in the plan that you can use to change for any of the weeks that don't align with you.

Breakdown of the routine

Repeat – Repeat each affirmation listed 10 times throughout each day phase. For example, let's take the morning affirmations for self-focus; I am happy, I am outstanding, I am excited to start my day. Repeat each affirmation in that order once, this is repeating each one of those a single time leaving nine more to go. It is preferred to chant the affirmations in this method as opposed to chanting one by one repeatedly. This allows a nice variation with each affirmation chanted sequentially.

Morning – The affirmations listed in the morning are to uplift your mood and affirm to yourself who you are. five times affirmation when you wake up, five times during your morning commute

Afternoon – Focusing on what you have and affirming your accomplishments. five times during lunch, five times for each during your commute.

Evening – Focusing on what you will do and affirming your actions. five times after dinner, five times before bed

Sequential or Single – You can either repeat the listed affirmations in one go or repeat each single affirmation the number of times recommended and move on to the next. For example, "I am happy, I am happy, I am happy" vs "I am happy, I am outstanding, I am excited to start my day. I am happy, I am outstanding, I am excited to start my day. I am happy, I am outstanding, I am excited to start my day." I prefer the ladder method but use what feels most comfortable for you.

Every affirmation you express needs emotions involved. Get into the habit of expressing affirmations with emotions. Usually, an emotion will be instantly involved with the chanted affirmation. If not, try a different affirmation and see how you feel.

Feel free to change any of the affirmations with ones more targeted or meaningful to you. Don't forget to download your copy of 1000+ affirmations found at the end of the book. Use any of the affirmations listed in this book and/or in the free eBook to target this plan for your goals.

Day #1-7: Self Focus

Morning:

- I am happy
- I am outstanding
- I am excited to start my day

Afternoon:

- I have an amazing life
- I have a life filled with great people
- I have endless possibilities for happiness

Evening:

- I will love life and live happily
- I will give love regularly
- I will wake up every morning with a smile

Day #8-14: Success Focus

Morning:

- I am successful
- I am driven to succeed
- I am a great leader

Afternoon:

- I have and provide superior service
- I have driven to push forward to success

- I have a great career

Evening:

- I will provide great value
- I will overcome any challenge
- I will live my ultimate dream

Day #15-21: Health Focus

Morning:

- I am healthy
- I am full of energy
- I am perfect, healthy, and whole

Afternoon:

- I have a healthy body and mind
- I have a fit and strong body
- I have endless health, energy, and vitality

Evening:

- I will live a long and healthy life
- I will keep my mind active
- I will bring energy with me everywhere I go

Day #22-30: Money Focus

Morning:

- I am rich
- I am dedicated to wealth
- I am increasing my net worth

Afternoon:

- I have all the money I want
- I have unlimited potential to achieve wealth
- I have clear sight of money-making opportunities

Evening:

- I will be positive about money
- I will acquire money easily and regularly
- I will amass wealth

Additional Focuses

Relationship Focus

Morning:

- I am a great partner
- I am loved by all who meet me
- I am focused on building healthy relationships

Afternoon:

- I have a loving companion
- I have the best of friends

- I have a warm and giving family

Evening:

- I will be loving to all
- I will build strong relationships
- I will always be supportive

Confidence Focus

Morning:

- I am confident
- I am self-disciplined and dependable
- I am an inspiration to others

Afternoon:

- I have confidence speaking my mind
- I have complete belief in myself
- I have unlimited determination

Evening:

- I will be confident in my daily activities
- I will move forward fearlessly
- I will control my circumstances

Advanced Methods

For those looking to increase the challenge of the 30-day plan, there are a few things you can add:

- Use visualization, with emotions, to put yourself where you want to be. Do this in the morning and at night before bed.

- Add inspirational quotes to your routine, morning, afternoon, or evening. Inspirational quotes are more effective if they come from someone who motivates or inspires you.

- Be grateful, wake up each morning and express gratitude for the things you already have. This is an excellent way to start your day and set your mind up for success in whichever focus you are on. This doesn't have to be exclusive to the morning, you can be grateful any time of the day.

- Write it down. Write down your gratefulness, affirmations, and quotes.

Advanced Plan

Plan

A single-focus plan is highly targeted at changing the way you think about a particular area in your life. This type of plan requires you to stay focused on one area during the 30 days. Use a similar routine with weekly splits but instead of a different focus for each routine, you have the same focus but different affirmations.

This is a great method to change your thoughts and beliefs about a particular area in your life. If you are new to affirmations, I suggest sticking to the plan laid out and then moving on to a targeted plan afterward.

Strategies

As you may have noticed, each day phase is specifically targeted with certain types of affirmations. The types of affirmations you use can be different and targeted towards your goals. I do recommend consistency during your 30 days of affirmations for the types you decide to use.

If your schedule is vastly different from most of ours, your plan should be adjusted accordingly. It doesn't matter when you chant the affirmations, just stick to a schedule that works for you and be consistent every day at those times.

Conclusion

The life you live is created first in your subconscious mind, these deep beliefs create your image, which creates your perspective, and your outer world reflects it to you. There is no separation between your mind, body, and the world you live in. Your body is physical and so is the world around you, therefore they are one thing. Both are an extension of the non-physical subconscious.

The piece of the puzzle most people are not seeing is their own piece and how they fit into this world. Most people never pay attention to their true desires, instead only focusing on what they should do to get through the day or pay their bills. Ignoring their dreams, ignoring their feelings, emotions and putting other people first to a fault.

Part of the reason for this is that for thousands of years the mass population was told that God and life were outside of them and that they are not worthy of having good things or to be wealthy. Instead we are all born in sin. We were taught that we need to be forgiven and to fear God. This made us neglect our internal reality because it was not worthy of our attention and our focus was directed only to the external world.

These concepts were taught to our ancestors which made us separated from reality and it has continued since then. These unworthy ideas sit deeply in the subconscious of the world. Any deeper connection to the world around us seems to not exist. To see yourself separated from the rest of reality is to neglect who you are on the inside, you will never see what shape your puzzle piece if you never look at yourself.

To learn to look within and recognize your connection to the outside world from the inside allows you to see who you are. How you fit in and how to live a happy, successful, and fulfilling life the way that is right for you. All the love you need and happiness you need comes from within you.

By looking outside for answers most people forget about themselves and therefore have a very low perspective of themselves. People don't feel worthy enough to attain the fulfillment of their dreams. From a young age, we are taught and shown that it is the things outside of us that make us happy.

We are not shown that our internal reality is more important and more real than the material world. Yet we strive to fulfill our internal happiness with external things. We were shown the concept of cause and effect, for example if you buy this new toy or attain that goal you will then be happy.

That type of happiness is always transitory because the external world is temporary and is always changing. We are taught to reach outside of ourselves, not realizing our perspectives on reality are backward from the real thing. You must first be happy to achieve your desires.

The truth of your being does not change and is not dependent on the external reality. By understanding that our true life is predominantly within us through our thoughts and feelings, we can learn to take control over those feelings and create a more lasting happiness.

As we have learned, polarities work both ways in life. A cause equals an effect, with practice and control of our thoughts and feelings we can create a good feeling first to attract more good things to us. Life is internal.

It is not just about taking the steps to fulfillment or success. it is about resonating with them first and then taking the steps from there. To be successful you must first feel and believe you are successful, you must know it in your heart, which is the center of all your feelings.

Your body, your life and the internal reality of your thoughts and feelings are your true gifts, the true you and it is more real than the outside world could ever be. Without you, everything you think is real does not exist.

It is your consciousness that makes your life a reality. Success and happiness stem from the inside, they are also skills that can be learned. Individually and as a world we must learn to live from our gut, and to see the world as our reflection, to see that we have a hand in creating our experiences.

We cannot change the outside world until we change ourselves. We cannot help others until we learn to help ourselves. We cannot help healing this world until we learn to heal ourselves. We cannot teach others what we have not yet learned.

Your world is literally in your head and heart, which is reflected at you. To change your world, you must learn to be at peace with what is and then think differently to spark the change of feeling and create new beliefs. After achieving a new mindset and new habits your reflection will change.

  This may take time and patience. Practice and repetition are your friends; they help you create a new world. A world that you call into existence from first believing in it. Through belief you will then see the manifestation and then with seeing your belief will be reinforced. This is something you do already, it is just that you haven't observed yourself enough to see it yet and you don't do it with purpose.

  Self-observation and noticing how your mind works will allow you to live more fully with purpose. The average person is generally not positive in their thinking and it is human nature to be lost in thoughts, to be lost in our imagination. The only times we are not lost in our imagination is when focusing on a task, when we solve problems or enjoy some part of our physical life like watching a great movie or playing a game.

Otherwise in the times between, like driving a car or doing some mundane activity we all live in our imagination. It is our default to worry about the past, worry about the future, be judgmental of others or situations. We relive scenarios or make them up by speeding up and needing to win arguments in our minds etc.

When we can observe ourselves naturally being taken away by our imagination, we can much more easily understand the power our minds have over us. The imagination is nothing to be trifled with. We have a will to choose better thoughts, this is our freedom of choice. To choose to envision the dreams and goals that we want, to literally dream them and feel them come to life with emotion.

Learn to watch how often you enter your imagination and listen to how powerful it feels when you get mad at someone or some situation that is out of your control. It feels so real.

This recognition allows you to learn that you already constantly use your imagination and allows you to choose better thoughts on purpose. Learn to take away your negative thoughts and replace them with your goals, think of them as often as possible, reroute your thoughts to be positive. Live in a positive, goal-oriented imagination as much as possible, using it on purpose can wipe out most negative thoughts and make your days happier. The goal is to make the imagination work for us not against us.

The success of your business depends on the vision you have of it and the steps you take to realize that vision. So, ask yourself what is "the end" result of having a successful business how should you act, how should you think? Regarding everything I have learned and how far I have come. I have come to the following conclusion and I try to live up to it as best as I can.

Having a successful business is more about the people you serve than anything else. They are the cause and the reason for the existence of the business. Therefore, to be successful means to look out for the clients own good irrespective of the solution being profitable to you. If you cannot help them solve their problems, then what you must do is redirect them to where they can find their solution.

You must learn to be honest and know when to step back and realize your purpose in this person's life is to help them find their solution with or without you. They will appreciate you just as much as if you did help them with your services and tell others about you. No matter what business you're in if it deals with money it deals with people.

Therefore, no matter what your business is, it is to serve others and that is exactly how you must operate to grow. The more of a people person you become, the more people you can serve, therefore the more profitable you become. Money is the side effect.

Money doesn't matter, it's the vision that matters. It's about the journey towards fulfilling your dreams.

To truly serve and to thrive and have a job that is not only financially successful but fulfilling also you need to be original. To be yourself, to bring your personality into work, to bring what you love into your work. To be able to be yourself in your work will give you passion to grow your business so you can do more of what you love. If you can love your job, people will come to you specifically because they can feel your confidence, they can feel your passion.

If you put yourself in "the end" result, how would you act and think if the situation you desired to manifest was already true. You must act the part. From a realistic point of view, you must try to live in that world as much as possible. Think, feel, and speak successfully.

Things just don't happen for any reason, first you resonate with them, to think from "the end", to live in "the end" is to create them. You must live and breathe the idea of success. You must focus your thoughts in the direction of fulfillment, toward the goal. You have the power of creative thinking and most people spend their days worrying about this or that, never choosing to focus in a positive forward direction.

Thoughts have power over your feelings, random worries can throw you into a whirlwind of stress. Learn to change your focus from worry to fulfillment of your goals. As you gain control of your focus, your subconscious will start to change, then your actions will start to change. As your actions change to fit your desires, you will see your path to success start to unfold in front of you.

True faith is keeping your goal in mind in your internal reality and taking the appropriate steps to get there. As you believe and move forward the world around you will adapt to accompany that belief. Use affirmations regularly, use daily meditations to focus on the end goal.

Just like you need to eat, sleep, and breathe, you need to daily focus on your goals with feeling. You need to take regular action steps to learn more, to be more. As you grow, your business will grow. Keep the target in mind and walk toward it.

You must be the change you want to see in your world. The center of your being is your heart and mind, with the right focus your life and the success of your business has no choice but to propel forward, just as the sun and the planets move forward through space. The key is learning how to focus your attention forward and the key to doing that is through repetition and continually reminding yourself how life works.

Meditation is one of the most helpful exercises you can ever do because it allows you to access to your subconscious, it allows you to observe who you are, how you think, what you want out of life and how to figure out how to get there. Coupled with the relaxing effect of meditation it is the most common activity successful people across the globe practice.

I cannot stress enough the importance of repetition and purposely taking time to focus on your goals. Your mind must be understood, for the power it holds over your life is great and you must make it work for you.

To help guide you forward in space and time to take you where you are focusing. You are the Law of Attraction, you are quantum physics, you are the atom and the electron, you are your goals, you are what you think and feel the most.

Choose to see your perspective over the reality around you until the reality around you changes to match it. You must have conviction, to have mind over matter, to believe in yourself and your dreams. If a situation is stressful choose to see it differently, superimpose a new perspective over it or don't think of it at all.

When you truly change your perspective and change your mindset the world will show you how you have changed but only when your subconscious mind has accepted the change.

I've heard that an airline pilot needs to navigate the plane 95% of the time, 5% of the time he or she doesn't have to. Just like life we need to continually refocus our attention on our desires and outcomes to achieve them. This is also like meditation practices where we need to continual re-focus our thoughts and breathing to achieve a deep meditative state.

You and your business are one, it is your baby, you must take care of it as such. Think good thoughts about it, continually be thankful for it and imagine it growing up. Through planning take the appropriate action steps to help it grow. Set goals and timelines to achieve them. Listen to your thoughts, listen to how you speak to others about yourself and your business.

You project your internal reality, so do your best to keep your thoughts and words in check. Doing this will be causing a new cause, creating a new effect. Your subconscious is reflected in the world around you so as you focus on growing your world will gradually show you the fruits of your labor. Learn to have patience, learn to love yourself more, learn to love thinking, learn to love the journey, and acknowledge the power of your mind by becoming the visionary of your life.

Samantha Goleman

CPSIA information can be obtained
at www.ICGtesting.com
Printed in the USA
BVHW091807190521
607714BV00003B/598

9 781801 828291

Quality Service, Competitive Business: setting the standard in customer service